Selling Science

Critical Issues in Health and Medicine

Edited by Rima D. Apple, University of Wisconsin–Madison, and Janet Golden, Rutgers University, Camden

Growing criticism of the U.S. health care system is coming from consumers, politicians, the media, activists, and healthcare professionals. Critical Issues in Health and Medicine is a collection of books that explores these contemporary dilemmas from a variety of perspectives, among them political, legal, historical, sociological, and comparative, and with attention to crucial dimensions such as race, gender, ethnicity, sexuality, and culture.

For a list of titles in the series, see the last page of the book.

Selling Science

Polio and the Promise of Gamma Globulin

Stephen E. Mawdsley

Rutgers University Press

New Brunswick, New Jersey, and London

Library of Congress Cataloging-in-Publication Data
Names: Mawdsley, Stephen E.
Title: Selling science : polio and the promise of gamma globulin / Stephen E. Mawdsley.
Description: New Brunswick, New Jersey : Rutgers University Press, 2016. | Series: Critical issues in health and medicine | Includes bibliographical references and index.
Identifiers: LCCN 2015037355 | ISBN 9780813574394 (hardcover : alkaline paper) | ISBN 9780813574400 (ePub) | ISBN 9780813574417 (Web PDF)
Subjects: LCSH: Poliomyelitis—United States—Prevention—History—20th century. | Poliomyelitis—Research—United States—History—20th century. | Gamma globulins—Research—United States—History—20th century. | Children—Diseases—United States—Prevention—History—20th century. | Hammon, William McD. (William McDowell), 1904–1989. | National Foundation for Infantile Paralysis—History. | Clinical trials—United States—History—20th century. | Science—Social aspects—History—20th century. | Science—Economic aspects—History—20th century.
Classification: LCC RA644.P9 M39 2016 | DDC 614.5/490973—dc23
LC record available at http://lccn.loc.gov/2015037355

A British Cataloging-in-Publication record for this book is available from the British Library.

Visit our website: http://rutgerspress.rutgers.edu

Manufactured in the United States of America

To my father, Robert L. Mawdsley

Contents

Illustrations

Acknowledgments

This book is the product of not only my own efforts and good fortune, but also the support of many kind and intelligent people. I am grateful for the advice and encouragement of historians Tony Badger, Angela N. H. Creager, Christopher Crenner, Helen Curry, Nick Hopwood, Joel Isaac, Judith Walzer Leavitt, M. Susan Lindee, Iwan Morgan, Andrew Preston, Patricia Prestwich, Leslie J. Reagan, Jonathan Reinarz, Naomi Rogers, Jane S. Smith, Matthew Smith, Susan L. Smith, and Heather Green Wooten.

I was privileged to correspond with retired polio researchers, who shared their recollections of the gamma globulin study. Thanks to Arthur E. Greene, Stanley A. Plotkin, Hilary Koprowski, and Julius S. Youngner. I also appreciate the assistance of Peter L. Salk in granting me access to the Jonas E. Salk archival collection and to Bill Kumm for supplying memories of his late father Henry W. Kumm. I am indebted to infectious diseases researcher Charles R. Rinaldo for his encouragement and to Coriell Institute executive Joe Mintzer for sharing his memories of Lewis L. Coriell.

The commitment of knowledgeable librarians and archivists was important to the realization of this book. I am grateful to David W. Rose, Roy Goodman, Charles B. Greifenstein, Lydia Vazquez-Rivera, Karen Jania, Malgosia Myc, Lynda Claassen, Stephanie L. Moll Bricking, Erik Moore, Nancy F. Lyon, David Mook, Kristen Rowley, and Tab Lewis. Moreover, genealogists Gloria Russell and Cliff Hayes were very helpful in recommending relevant newspaper collections.

The assistance of journalists in Utah, Texas, and Iowa, helped me to connect with former child participants and their families. Thank you to Ace Stryker, Cody Clark, Steve Jetton, John Wilburn, and Joanne Fox.

This book was generously funded through a combination of fellowships, scholarships, and grant schemes. I acknowledge the support of the Government of Canada's Social Sciences and Humanities Research Council and the Government of Alberta's Lougheed Award. I am also thankful for England's Overseas Research Studentship from the Higher Education Funding Council and the Canada Cambridge Scholarship from the Cambridge Commonwealth Trust. Equipment and conference costs were deferred by Clare Hall's graduate award and research schemes. Further support came by way of Cambridge Faculty of History grants, including the Sara Norton Fund, the Members History

Fund, the Chair's Fund, and the Hannay Doctoral Training Fund. In addition, the Cambridge Department of History and Philosophy of Science supported my research through the Williamson and Rausing Trust. I am thankful to my college, Clare Hall, for awarding me the Isaac Newton–Ann Johnston Research Fellowship, which gave me the resources and time to complete this manuscript.

Finally, I am indebted to my partner, Helen Mawdsley, whose endless patience, enthusiasm, and constructive criticism helped me finish this book.

Abbreviations

American Journal of Public Health	*AJPH*
American Medical Association	AMA
American National Red Cross	ARC
Church of Jesus Christ of Latter Day Saints	LDS
Communicable Disease Center	CDC
Gamma Globulin	GG
Institutional Review Board	IRB
Joint Orthopedic Nursing Advisory Service	JONAS
Journal of the American Medical Association	*JAMA*
March of Dimes Archives	MDA
National Foundation for Infantile Paralysis	NFIP
National Research Council	NRC
Office of Defense Mobilization	ODM
United States Public Health Service	USPHS
Vaccine Evaluation Center	VEC
World Health Organization	WHO

Selling Science

Introduction

"There could be almost complete confidence that, if and when a [polio] vaccine [was] developed, the American people would back the scientific trials necessary to test its effectiveness."[1] This assertion was penned in 1956, at a time when the eradication of the fearsome disease, polio, was well under way. Funded by the National Foundation for Infantile Paralysis (NFIP), the vaccine developed by Dr. Jonas Salk and evaluated in a massive 1954 clinical trial was found to be safe and effective.[2] Such characterizations of public support for human medical experimentation were evidently assumed and linked to earlier developments. What had come before the polio vaccine trial to normalize enrolling millions of healthy children to test a new medical intervention? What role did publicity play in shaping perceptions of medical research? This book attempts to unravel these questions, while delving deeper into the nature of medical experimentation conducted on an open population in mid-twentieth century. At a time when most Americans trusted scientists and the NFIP, but knew no model for a mass clinical trial, their mutual encounter under the auspices of conquering disease was shaped by politics, marketing, and, at times, deception.

Poliomyelitis—or polio, as it became abbreviated—is a contagious oral-fecal viral disease.[3] Contaminated food, hands, and objects are the most common means of spreading the virus from person to person.[4] Smaller than bacteria, viruses such as polio are microscopic entities that were known to medical scientists before the 1890s but not viewable until the widespread use of the electron microscope during the 1940s.[5] Polio infects a living cell and uses it as a host in which to replicate until the viral copies rupture the cell membrane; the copies may then spread to new cells. There are three known serotypes of poliovirus, and an individual requires protective immune antibodies to each type to guard against illness. When the virus is ingested by a nonimmune individual, it can cause a gastrointestinal infection and produce symptoms resembling a mild flu.[6] All persons infected with the disease, irrespective of its severity, will pass the virus in their feces. Although most sufferers will recover from the intestinal infection and develop lasting immunity to the offending viral serotype, in approximately 5 to 10 percent of cases the poliovirus will pass into the bloodstream, where it targets the motor neurons of the spinal cord. As the attack progresses, lesions develop in the gray matter of the spinal cord, informing the Greek-derived term: *polios* (gray) *myelos* (marrow), *itis* (inflammation).[7] Although adults can become afflicted, polio holds a peculiar affinity for children, which inspired the clinical synonym *infantile paralysis.* Depending on the severity, location, and number of spinal lesions, patients may experience paralysis of the extremities and the respiratory muscles. Complications arising from severe infections can lead to death.[8]

When polio arose in epidemic form in America at the turn of the twentieth century, no means to prevent the disease existed, and its nature was poorly understood.[9] As the disease appeared to propagate during warm weather, summer soon became known as "polio season."[10] To thwart the spread of polio, public health departments closed areas where contagion appeared highest, such as playgrounds, cinemas, and swimming pools, and encouraged hygiene and fly eradication programs.[11] Even though public health efforts controlled other diseases, such as cholera and typhoid, polio did not respond to collective health activism and persisted even in salubrious neighborhoods.[12] Parents and health workers waited with trepidation for news of the next outbreak.

Health professionals and researchers struggled to understand what factors predisposed individuals to a polio infection.[13] Although flies and immigrants were blamed initially, shifts in scientific knowledge challenged assumptions.[14] Most scientists theorized that society's growing adherence to hygienic practices denied infants exposure to the virus while still protected by maternal antibodies.[15] Doctors could only counsel vigilant lifestyle choices for polio prevention: stay clean, avoid changes in temperature, get plenty of rest, and avoid crowded places. Out of desperation, citizens turned to quarantine to isolate the ill and regulate the regimens of the healthy.[16] Communities attempted, with limited success, to control this seemingly random affliction that defied public health ordinance.

A truly national effort to battle polio emerged through the vision of Franklin Delano Roosevelt.[17] In August 1921, thirty-nine-year-old Roosevelt was stricken with an illness believed to be polio during a family vacation at

Campobello Island, New Brunswick, Canada.[18] Roosevelt's bout with the disease interrupted his political career and forced him to undergo a lengthy convalescence. He assessed a range of therapies to restore his paralyzed leg muscles, but with minimal effect. On the suggestion of a close family friend, Roosevelt traveled in 1924 to Warm Springs, Georgia, in search of the purported curative powers attributed to warm mineral water pools. After Roosevelt began to benefit from the rustic Georgia environs and warm water swims, he purchased the dilapidated Warm Springs property in April 1926 for $201,677 and turned it into a treatment resort for polio survivors.[19] Although he never regained use of his leg muscles, Roosevelt believed that the facility could help other polio survivors and serve as a potent symbol of hope.[20]

As patients flocked to Warm Springs seeking miracle cures, the financial burden of building maintenance and medical attendants became untenable. With the assistance of his law partner, Basil O'Connor, Roosevelt formed the Georgia Warm Springs Foundation in July 1927 to raise donations from Democratic Party supporters and fund the growing polio treatment program. When Roosevelt was elected president of the United States in 1932, the skillful public relations efforts of Carl Byoir expanded the polio fund-raising campaign. Byoir advised Roosevelt supporters to organize Presidential Birthday Balls under the motto "Dance so that others may walk."[21] The funds collected from these annual ticketed events across hundreds of communities kept Warm Springs financially viable during the Great Depression and situated Roosevelt as the first celebrity patron of the polio crusade.

By 1937, controversy over Roosevelt's economic policies and the political nature of the Presidential Birthday Balls forced a reassessment of the program. To extricate the polio crusade from the quagmire of federal politics and broaden charitable appeal beyond the Democratic Party, O'Connor and Roosevelt astutely conceived of an ostensibly nonpartisan charitable institution. From his law offices in New York City, Basil O'Connor incorporated the National Foundation for Infantile Paralysis (NFIP) on January 3, 1938.[22] The mandate of this new charity was to raise money from Americans to fight polio by investing in medical treatment, health education, and scientific research. The NFIP program offered a coordinated response to polio and hope that the disease would one day be eradicated.[23]

The NFIP was directed by a board of trustees comprised of prominent American business leaders, such as the president of International Business Machines, Thomas J. Watson, but the majority of operational power was entrusted to its president, Basil O'Connor.[24] Born on January 8, 1892, in Taunton, Massachusetts, O'Connor was the son of working-class Irish immigrants. Motivated to transcend his father's trade as a tinsmith, he channeled his energies into academic pursuits and graduated in 1912 from Dartmouth College in Hanover, New Hampshire. His outstanding academic performance and connections with influential alumni facilitated his entrance into Harvard Law School. Upon completing legal studies in 1919, O'Connor relocated to New York City, where he fortuitously befriended Roosevelt and together they established a Wall Street corporate law firm. The economic boom of the 1920s combined with Roosevelt's prominent association enabled O'Connor to amass considerable wealth and enjoy the trappings of economic success, including "private railroad cars, limousines, a gentleman's farm on Long Island, and a ready table at the most expensive restaurants."[25] During the Great Depression, O'Connor served as a personal adviser to Roosevelt and helped in the recruitment of corporate leaders to the Brain Trust and Cabinet.[26] With the founding of the NFIP, O'Connor used the organization as a platform to exercise influence and realize his desire for fame; he demanded high standards for the charity and nurtured its brand as America's foremost polio philanthropy.

As a political insider and lawyer versed in the strategies behind prosperous capitalist enterprises, O'Connor modeled the NFIP around a hierarchical structure.[27] He established several specialist departments attuned to specific institutional needs, such as public education, public relations, fund-raising, chapters, medical treatment, and medical research.[28] Beyond the paid staff at headquarters, O'Connor and the board of trustees built a remarkable volunteer network through affiliated county chapters. Operated by largely middle-class unpaid workers, the chapters implemented the philanthropic program at a local level. Chapters coordinated payment for polio hospitalization and treatment, distributed educational materials, and helped to organize fund-raising. Unlike the Rockefeller or Carnegie foundations, the NFIP was not bestowed with a large operational grant to maintain its program; consequently, successful fund-raising was vital for institutional survival. The March of Dimes, a play on words by comedian Eddie Cantor in reference to the "March of Time" newsreels, became the name of the NFIP fund-raising campaign. Launched every January, the March of Dimes was operated by volunteers and supported by headquarters publicity and training initiatives.[29] Half of the donations raised in each March of Dimes campaign were transferred to county chapters so that volunteers could pay for local treatment, while the remainder was retained at NFIP headquarters to maintain the national program.[30]

The rising incidence of polio meant that money for medical treatment was desperately needed. Since the symptoms of polio were varied, most doctors during the 1920s and 1930s confused the disease with other ailments, such

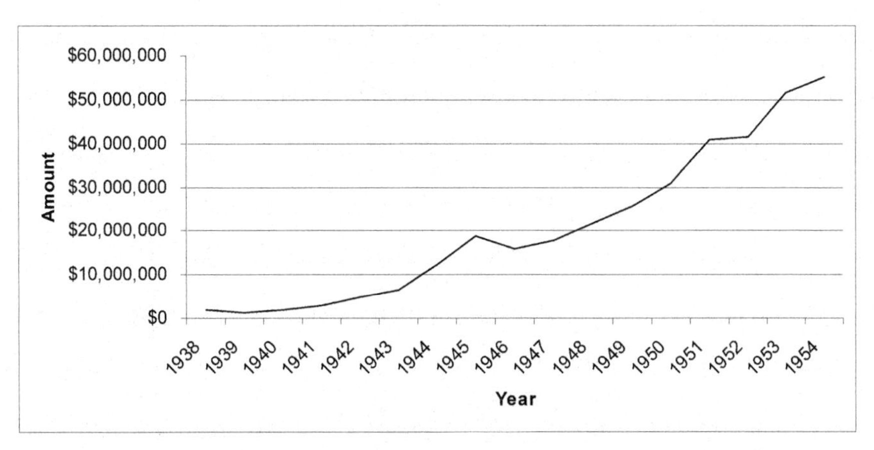

Figure 1 Contributions raised by the March of Dimes campaign, 1938–1954.

Source: Stephen E. Mawdsley, "Polio and Prejudice: Charles Hudson Bynum and the Racial Politics of the National Foundation for Infantile Paralysis, 1938–1954" (MA thesis, University of Alberta, 2008), table 3.

as influenza or encephalitis. Only when medical training and diagnostic procedures improved after World War II did previously hidden cases enter the record.[31] In 1938, fewer than two thousand cases of polio were reported; a decade later, rates of infection had climbed to over twenty-seven thousand cases. The endemism of polio combined with improvements in public health reporting increased awareness of the illness and pressure to respond.

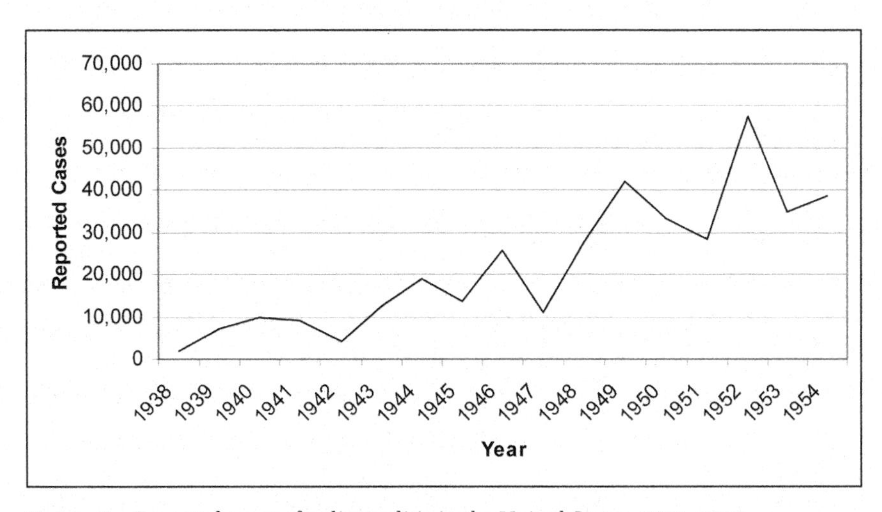

Figure 2 Reported cases of poliomyelitis in the United States, 1938–1954.

Source: Stephen E. Mawdsley, "Polio and Prejudice: Charles Hudson Bynum and the Racial Politics of the National Foundation for Infantile Paralysis, 1938–1954" (MA thesis, University of Alberta, 2008), table 2.

Medical treatments for polio attempted to address immediate and long-term needs. When an attending physician suspected a case of polio, the patient was usually taken to the nearest hospital or acute treatment facility for testing and observation. Under these uncertain conditions, many polio patients struggled to understand what was attacking their bodies. "I remember crying a lot because I was so frightened," one survivor recollected. "What is wrong with me? Was I dying? Is this something really bad, Mom? I didn't say the word either. I didn't say polio. It was a horrendous fear. I was just afraid I was going to die."[32] Once diagnosis was confirmed through a spinal tap or limb flexibility tests, acute care was instituted.

Acute polio care was conducted in hospital isolation wards and orchestrated around routines intended to save lives, restrict contagion, and reduce the extent of paralysis.[33] Although attended by nurses and physicians, polio patients were often overwhelmed by a sense of dread and loneliness. "The other children would scream for their parents all the time," recalled former patient Mark O'Brien, "especially in the evenings. They were either crying or yelling for their mommies and daddies. It was very hellish and scary."[34] The formidable pain associated with acute polio was notorious. During her bout with the illness, Joan Morris remembered, "I could not stand for anyone to touch me because the pain was so bad. . . . Just to touch my skin would cause me to scream in pain."[35] Specialized lifesaving equipment was often employed, including the large cylindrical "iron lung" breathing apparatus, which aided patients suffering from bulbar (respiratory) polio. The experience of being housed in a metal machine that would "just pump and hiss and gush and pump" was uniquely chilling for bulbar polio patients, many of whom were uncertain whether they would become well again.[36] Its predominant application in polio made the iron lung an iconic symbol, representing both the hope and the terror associated with the affliction.[37]

Polio treatment was transformed in the early 1940s when an Australian health activist and physical therapy pioneer, Sister Elizabeth Kenny, challenged medical orthodoxy. Before Kenny arrived, most American doctors considered polio a nerve disease and favored immobilization of paralyzed limbs with casts or splits. Kenny, however, theorized that polio was a systemic disease and rejected immobilization; she reasoned that paralysis must be treated early with exercise and hot compresses to reduce muscle spasms. "I remember the steamer kettles for the hot packs," patient Charlene Pugleasa recalled. "There would be one layer, wet wool, and then they would come with this thick, heavy plastic piece that was cut the same shape and they'd put that over it. . . . I loved the feeling because it was a comforting feeling."[38] Kenny's treatments gave patients hope and improved prognoses, but her lack of formal training combined with a forceful manner alienated many medical allies. Kenny was initially funded by grants from the NFIP, but rising professional resentment of her manner and method led to a break with her sponsor and inspired the founding of a competing charity and training center, the Kenny Foundation. Over time, many of Kenny's methods were found to be effective and subsumed into clinical practice, leading to important shifts in polio treatment.[39] Despite such innovations, most nurses and doctors had little choice but to wait until the acute stage passed. "There was no treatment other than supporting respiration, swallowing food and water, and the hot packs," retired physician Dr. John Affeldt explained.[40] Although some patients were treated outside institutional settings, hospital wards were the primary polio battlegrounds.

For most survivors of the acute infection, months or even years of convalescent treatment lay ahead at special rehabilitation facilities. While Warm Springs, Georgia, remained central to polio therapy, several other centers were established across the nation, including one for African Americans at Tuskegee, Alabama.[41] Some patients were determined to overcome their disability, while others struggled with the painful psychological and physical transition. "When I saw the other kids fitted with their braces and long crutches, I knew the same thing would probably happen to me," Tuskegee patient Clara Yelder recalled. "I had resigned myself that I wasn't going to walk again."[42] Many paralyzed survivors endured arduous physiotherapy exercises and painful orthopedic surgeries to correct muscle atrophies and regain lost mobility. "I underwent surgery that included a spine fusion of my lower back with a tibual graft," patient Robert Huse explained. After successive surgeries and joint "manipulation," he was discharged to his family.[43] Like Huse, nearly all survivors greeted the move from institutional to home care with relief. "I remember the day I came home," explained Mark O'Brien. "My father had installed an intercom to connect my room with my parents' room. . . . I was very happy to get home."[44]

The need for special equipment and trained personnel made polio an expensive disease to treat. The combined cost for acute and convalescent care frequently reached into the thousands of dollars per person. Roosevelt's New Deal initiatives, including the Social Security Act of 1935, provided states with economic resources to establish services for "crippled children," but the quality and degree of such offerings varied by region.[45] Although health charities, such as the NFIP, complemented state programs by covering acute and convalescent treatment costs, not all families had access to or sought such aid. Joan Morris recalled how her parents experienced difficulties contending with the costs. "Mom went to work at the Shade and Curtain Company," she remembered.

"She started to work the first day after I went into hospital. She made $15.00 a week. $14.00 went for my care, and $1.00 went to pay for her lunches and to pay for the streetcar ride back and forth."[46] Many layers of financial support were often necessary to subsidize the expense of polio treatment.

Although a few patients achieved near-complete recovery, most experienced a legacy of lasting paralysis.[47] Polio patient Mark O'Brien spent the majority of his life in a respirator. "The doctor said I should be back in the iron lung, and so they delivered an iron lung to our house," O'Brien recalled. "It just barely fit in my room."[48] Many others required the assistance of crutches, wheelchairs, or leg braces. Robert Huse remembered how his dependence on crutches affected his mobility at school. "I soon discovered," he explained, "that although I could navigate all the stairs, the great distances that must be traversed after each class resulted in my being consistently late for each one."[49] Since the extent of recovery varied, polio survivors adapted to their circumstances.[50]

The social stigma of disability created challenges for many polio survivors.[51] "With the reality of being crippled," Morris reflected, "a child either withdraws from the risks of socialization or finds ways to cope and defend itself out in a world."[52] For some children, the prospect of returning to school and rebuilding friendships was fraught with anxiety. Physical education was one part of the curriculum that could discourage survivors. "When the kids in my class took gym class," Joan Morris explained, "all I could do was sit and watch them. Believe me that was not much fun."[53] Finding paid work was also difficult for survivors, since most employers considered physical disability an impediment to productivity.[54] Even though some survivors found occupations that accommodated disability, they and their families negotiated a society discomforted by the remnants of polio.[55]

During outbreaks, fear of contagion often provoked defensive behavior. Parents restricted the social activities of their children and were advised to avoid interacting with new groups. Restaurants closed and popular events were cancelled. "At the height of the epidemic," Dr. Richard Aldrich recollected, "the people of Minneapolis were so frightened that there was nobody in the restaurants. There was practically no traffic, the stores were empty. It just was considered a feat of bravado almost to go out and mingle in the public."[56] For wage earners, the disruption was more than an inconvenience. Pugleasa recalled how her father's work experience changed once colleagues learned of his daughter's illness. "My father also had to take a shower as soon as he got to the mine, leave his street clothes in a bag, so as not to infect the other men before he put on his mining clothes," she explained. "It was that shower before going down that made him feel like he was separated."[57]

Some families with stricken children were ostracized by neighbors who were concerned that any interaction might spread the disease.[58] Other people imagined polio as a form of divine retribution, believing parents were punished with polio because they were not "churchgoers."[59] Many communities became divided and suspicions flourished. In fact, concern over stigmatization led some parents to send their paralyzed children to institutions, rather than risk becoming social pariahs.[60] Still others were reluctant to take on the added responsibility of home care. A former patient recalled that when she wanted to return home, her mother became angry. "My mother was very upset and believed that I was manipulating. . . . And it hurt me terribly . . . because she said, 'No, we're not coming to get you. You are better off there. They're helping you. We're not coming to get you.'"[61] Fear of contagion tested the bonds of communities and families alike.

Racial prejudice further complicated responses to polio. Although the NFIP treatment program was purportedly available to all Americans irrespective of "race, creed, or color," realizing such tenets was difficult and often unsuccessful. During the 1940s and 1950s, America was divided along racial lines, with legalized segregation in the South and de facto segregation in the North. Although the civil rights movement was gaining momentum, the notion of "separate but equal" facilities, as articulated in the 1896 *Plessy v. Ferguson* Supreme Court decision, remained in force.[62] Most hospital wards were segregated, and treatment for blacks rarely matched that of whites.[63] Like tuberculosis or syphilis, polio was steeped in a strong racial association: it was widely considered predominantly an affliction of white Americans.[64] Sophisticated March of Dimes publicity programs aimed at white middle-class donors, combined with prejudiced public health reporting and restricted access to medical care kept countless black polio cases hidden.[65] It was not until the 1940s that African Americans were recognized as legitimate polio survivors; however, such awareness was constrained by region and urban segregation.[66] Although most polio researchers and NFIP officials agreed that race was not a factor in polio susceptibility, they nevertheless saw whites as the most important contributors to and beneficiaries of their program.

Perceptions of and reactions to polio were shaped by the media. Newspaper writers brought to readers emotional accounts of desperate families, theories of causation, and alarming statistics about the growing rate of infection. "Polio Cases in U.S. Up 71% This Year" declared one 1940s *New York Times* headline.

The accompanying article quoted an NFIP official, who explained somewhat apologetically, "Frankly, we had predicted light incidence . . . because in the previous three years the incidence in the large population centers was high, and we thought the susceptible population was small."[67] The names and addresses of stricken individuals were listed in newspapers, exacerbating the culture of suspicion already evident during epidemics. The media not only captured the sense of powerlessness, but also helped make polio and its consequences front-page news.

Complementing media coverage, the NFIP devoted substantial resources to public education and March of Dimes publicity.[68] Epidemic-awareness pamphlets were widely distributed through the chapter network and circulated to schoolchildren and their parents. Foundation news releases were regularly sent through the wire services, offering the latest scientific information or advice about infection. Beginning in 1946, photogenic child polio survivors were selected to adorn national fund-raising posters.[69] These "poster children," frequently depicted in wheelchairs and on crutches, were potent reminders of polio paralysis. Short films about polio and the March of Dimes, such as *The Crippler*, were introduced during the 1940s to headline some cinematic productions.[70] One reporter offered this description of *The Crippler*: "A dark cloud spreading over playground and farm, mansion and tenement, a cloud that takes the shape of a hunched and sinister figure who cackles over his many victims."[71] The dramatic approach taken by NFIP producers heightened the fear of polio, but also delivered a potent message of hope through charitable donation. "We went to the theater often, we didn't have television," remembered one polio survivor. "And before the movie there was always the newsreel, which carried with it this great big thing about polio and the epidemic. And it kept us abreast of what was going on, and it kept everybody very afraid."[72] As television became increasingly popular by the 1950s, the NFIP organized special March of Dimes broadcasts featuring celebrities, including Louis Armstrong, Sammy Davis Jr., and Lucille Ball. The public relations arm of the NFIP helped increase awareness of polio, while also presenting the charity as the best hope to fight the disease.[73]

Knowing that no amount of treatment would conquer polio, Basil O'Connor established advisory committees to steer the long-term aims of the charity.[74] With initial input from microbiologist Dr. Paul de Kruif and under the guidance of former Rockefeller Institute research director, Dr. Thomas M. Rivers, the NFIP set up a special committee on medical research. During the 1930s and 1940s, when medical research grants were paltry and scarce, the NFIP research committee generously funded investigations in epidemiology, immunology,

Figure 3 March of Dimes poster child Nancy Drury, 1947. Courtesy of the March of Dimes Foundation.

and virology. Nearly $69 million was awarded by the NFIP between 1938 and 1962.[75] Although federal government investment in medical research increased after World War II, the expansive patronage network established by the NFIP provided the framework for basic and applied polio research.[76] Dr. John R. Paul, a medical researcher at Yale University, once recounted that the NFIP's support of science was akin to "the sudden appearance of a fairy godmother of quite mammoth proportions who thrived on publicity."[77]

Since the safety and efficacy of new medical interventions needed to be evaluated, clinical trials were conducted. These trials usually followed animal testing and were "diagnostic, preventative, or therapeutic in nature." Most trials adhered to a protocol, which defined the study's methods, organization, statistical underpinnings, and objective.[78] Prior to the establishment of Institutional Review Boards in the 1970s for ethical oversight, the medical researcher or chief investigator was ultimately responsible for assessing health risks and shouldering outcomes. Human subjects were drawn from either open or institutionalized populations (known as healthy volunteers/subjects) or from hospitalized populations (known as patient volunteers/subjects).[79]

Clinical trials were and continue to be used to assess immunizing agents for disease prevention. There are two forms of immunization: active and passive. Active immunization is characterized by introducing a killed or attenuated disease agent, such as a virus or microbe, into an individual as a means of prompting the immune system to develop antibodies for lasting protection. Passive immunization is achieved when antibodies from an immune individual are harvested from blood and transferred by injection to a nonimmune individual for temporary protection.[80] Since all clinical trials harbor risks and many require submitting to unpleasant procedures, human subjects and their guardians require some form of persuasion to volunteer; scrutinizing the process of persuasion reveals the negotiation between researchers and human subjects.[81]

This book examines the first large clinical trial to control polio using healthy children drawn from an open population. In the early 1950s, medical researcher Dr. William McD. Hammon and the NFIP launched a pioneering medical experiment on a scale never witnessed before. Conducted on over fifty-five thousand children in Utah, Texas, Iowa, and Nebraska, the study aimed to assess the safety and efficacy of gamma globulin (GG) in preventing paralytic polio. Although the study was condemned by many health professionals, harbored health risks, and returned dubious results, it was hailed as a triumph and used to rationalize a federally sanctioned mass immunization study on thousands of families between 1953 and 1954. Far from being transparent, the concept, conduct, and outcome of the GG study were sold to health professionals, medical researchers, and the public at each stage. When evidence revealed that GG did little to control polio, a publicity program was unleashed in defense. At a time when most Americans trusted scientists and the NFIP, contrived publicity and fear of contagion enticed them into an unequal alliance with science.

In spite of a growing body of scholarship on polio in the United States, historians have not investigated the testing of GG or how its use reveals the nature of clinical trial marketing during the Cold War.[82] Unlike scholarship exploring the enrollment of institutionalized or marginalized populations in research, this book examines how white and African American parents were persuaded to volunteer their healthy children for research. Before the GG field trials, scientists did not know whether the public would willingly participate in experiments that did not guarantee safety or efficacy. Medical researchers' indifference toward medical ethics led to a program that held parallels with the Tuskegee syphilis study and later polio vaccine trials.[83] By drawing on oral history interviews, medical journals, newspapers, meeting minutes, and private institutional records, the book explores the effects of marketing on medical research and the legacy of the gamma globulin clinical trials.

Chapter 1

Forging Momentum

In the closing days of World War II, the rural Catholic orphanage of Saint Vincent's near Freeport, Illinois, was mobilized for a medical experiment. A polio outbreak had taken hold in the facility, and there were many cases of paralysis. On the morning of August 26, 1945, a team of public health officers and scientific consultants arrived at Saint Vincent's, seeking to control the outbreak by injecting a human blood fraction as part of a clinical trial.[1] That afternoon, Saint Vincent's staff divided the young residents along age and gender lines in preparation for injections. An efficient, assembly-line approach characterized the experiment, as a team of two nurses prepared syringes while two physicians administered the injections. That day, Saint Vincent's charges became among the first humans to systematically receive gamma globulin (GG) in the battle against polio.[2] Why did researchers believe that the blood fraction might be effective against the disease? What factors complicated the testing of GG in humans? This chapter explores the discovery of the blood fraction and how medical researchers attempted to assess its value for polio control.

Gamma Globulin in the Fight against Disease

The entry of GG into the lexicon of clinical interventions was a not radical departure, but was freighted on a long-standing fascination with human blood. Since the early twentieth century, doctors had believed that blood harbored protective properties that could be conveyed to others through injection. Some physicians experimented with blood to treat measles and scarlet fever with some success. As a result, the idea that the curative qualities in blood could be passed on to vulnerable individuals lingered, awaiting new studies and technology.[3]

Laboratory breakthroughs during the late 1930s and early 1940s rekindled interest in human blood to fight disease. Dr. Charles Armstrong's development of an animal model for polio provided researchers an inexpensive means to evaluate new medical interventions.[4] In addition, the 1936 discovery of human blood fractions connected prior optimism with a clinical reality. Dr. Arne Tiselius, a chemist at Uppsala University, Sweden, designated γ (gamma) globulin to a group of unique blood proteins.[5] Tiselius found that protective proteins, called antibodies, permeated this fraction, but it was difficult to extract and concentrate them for clinical application.[6] American researchers, funded through federal government initiatives ahead of World War II, turned Tiselius's breakthrough into a practical solution. In 1940, the surgeon general of the U.S. Army solicited the National Research Council (NRC) to assist in meeting the military's medical needs for chemotherapy and blood transfusions.[7] The NRC established a series of committees that concerned themselves with the development of blood products, which could be stockpiled for use in distant theaters.[8] Research funding combined with the impetus of impending conflict initiated a program to unlock the secrets of blood.

To investigate methods for processing and stabilizing blood products, the NRC turned to Harvard physical chemistry professor Dr. Edwin J. Cohn.[9] Born to a family of wealthy New York City tobacco merchants in 1893, Cohn was educated at Amherst College, earned a degree from the University of Chicago, and pursued graduate research at Yale and Harvard.[10] By 1940, he was director of the Department of Physical Chemistry at the Harvard Medical School, where he and his team worked on proteins and blood albumin.[11] Applying this collective knowledge of proteins, peptides, and amino acids to blood, Cohn's team devised a novel ethanol and centrifugation technique that separated blood into its five constituent parts—or fractions. Not until Cohn's fractionation process was wedded to the American National Red Cross (ARC) blood donor program, however, were clinical possibilities brought to fruition.[12] The result was a range of new blood products, including fibrogen (a clotting agent), gamma globulin (an antibody solution), and serum albumin (the purified liquid component of blood). When serum albumin was successfully administered in December 1941 to treat shock among Allied personnel at Pearl Harbor, the blood fractionation program was elevated to a pillar of the war effort. The U.S. Navy, enthralled with the potential medical value of blood fractions, encouraged their clinical application and contracted American pharmaceutical suppliers, such as Sharp & Dohme and Squibb & Son Co., to provide a steady supply.[13] Health professionals and researchers clamored to assess the efficacy of gamma globulin in preventing illness. Although GG, as a derivative of blood, was not homogeneous and could vary in potency based on the size of the donor pool, its concentrated antibody composition promised remarkable benefits.

Among the first researchers to evaluate GG and determine its clinical utility was Dr. Joseph Stokes Jr. Born in 1896 in Moorestown, New Jersey, into a family of prominent physicians, Stokes attended Haverford College, earning a BA in 1916 and an MD in 1920 from the University of Pennsylvania Medical School.[14] After interning at the Massachusetts General Hospital in Boston, Stokes returned to Pennsylvania, where he specialized in pediatrics at the Children's Hospital of Philadelphia. Between 1923 and 1931, Stokes held various teaching positions at the Penn Medical School, where he earned full professorship and became chief of pediatrics.[15] By 1942, he was appointed director of the Measles and Mumps Commission with the Armed Forces Epidemiology Board, while also acting as a consultant for the surgeon general.[16] In these assignments, he assessed the potential value of GG in controlling influenza, measles, mumps, and hepatitis.[17] During an outbreak of hepatitis at a summer camp near Philadelphia, Stokes administered GG and discovered that while 67 percent of the non-inoculated campers became ill, over 80 percent of the GG recipients remained well.[18] Stokes also injected GG to control measles, which reportedly "prevented attacks" or "made them mild."[19] Through these studies, Stokes showed that GG had important clinical applications.

Complementing Stokes's research, Dr. Sidney D. Kramer, the associate director of the Michigan State Department of Health, experimented with GG for polio. Kramer had a long-standing interest in blood products and developed a prototype polio vaccine that combined convalescent serum with live virus.[20] In 1943, Kramer began experiments with monkeys and found that GG provided protection when injected eight to sixteen hours before a challenge response with poliovirus.[21] Encouraged by these findings, Kramer believed that a human study was justified. However, acquiring enough GG for a clinical trial was difficult during World War II, as it was in short supply and closely regulated by the ARC and NRC. Since Kramer was doubtful that he could successfully obtain approval alone, he turned to the NFIP as a sponsor and negotiator.

NFIP medical director Dr. Donald W. Gudakunst was eager to support Kramer's plan. Born in 1895 in Paulding, Ohio, Gudakunst earned a medical degree from the University of Michigan in 1919 and developed a strong background in public health during the 1930s as Detroit's deputy health commissioner and later Michigan State health commissioner.[22] Gudakunst was interested in pragmatic solutions to polio and in 1943 arranged for the NFIP to sponsor Kramer's GG clinical trial.[23] Since paralytic polio was relatively rare compared with other illnesses, Kramer estimated that hundreds of children would be needed to assess the potential value of the blood fraction. Kramer and Gudakunst collaborated on a protocol and submitted it to the ARC and NRC for consideration. However, when a severe epidemic emerged in Chicago, Illinois, Gudakunst was surprised to learn that both national agencies rejected the plan and would not release the blood fraction for the human study.[24] Despite disappointment, Gudakunst and Kramer worked together to build support for a second application.

Many leading American medical scientists were in favor of testing GG as a means to control paralytic polio. In May 1944, researchers attending the NRC conference in Washington, D.C., discussed recent scientific developments, including the clinical use of blood fractions. A shared opinion emerged from this meeting that "the usefulness of human globulin concentrates on the human [polio] infection should be explored."[25] Dr. Joseph Stokes added to the momentum by writing to Gudakunst and Cohn, explaining that he was "considerably disturbed" that there were no known plans to assess GG for polio.[26] The most "common sense" approach, Stokes argued, was to establish a NFIP committee to monitor polio epidemics and select a high-incidence community to serve as a test site. Stokes invoked the emotive call to action, long attributed to the March of Dimes fund-raising campaign, by asserting that "if our own children were in a summer camp or in a school where a sharp outbreak occurred, we would think of gamma globulin first and would probably use it."[27]

Stokes offered advice on how NFIP officials might gain support from the ARC. He advised that any future study be "jointly supervise[d]" by the ARC and the NFIP. He advocated a reciprocal arrangement between trial participants and scientific investigators, so that for every 20 cc injection of GG into a child, the parent would be asked to donate one pint of blood. Stokes believed that parental blood donations would not only supplement ARC supplies, but also provide high-quality blood rich in polio antibodies. He hoped that ARC officials would look favorably on any program that increased blood supplies at a time of war.[28] To bolster the scientific merit of the trial, Stokes advised observed controls. His rationale for selecting the control group was based on the tension between public faith in science and the fear of health risks. He reasoned that when some parents learned that the study was "experimental," they would "not accept the procedure" and by default become the "essential" observed control.[29] Stokes believed that fear of experimentation could be harnessed for scientific benefit. Although he did not offer to conduct the study himself, he added conviction, advice, and impetus to proceed.

Like Stokes, federal government and military researchers advocated a GG study. In July 1944, Dr. Alphonse Raymond Dochez at the Office of Scientific

Research and Development encouraged Kramer to submit a clinical trial protocol to the ARC.[30] Likewise, Dr. John H. Dingle at the Commission on Acute Respiratory Diseases at Fort Bragg, North Carolina, spoke in favor of a GG clinical trial.[31] Dingle recognized the challenges such a study posed, since "more than two injections" per person would be needed to outlast an epidemic. He further postulated that urban centers were most suitable as test sites due to their "more consistent epidemic behavior, better diagnostic facilities, [and] geographic concentration."[32] In the waning years of the World War II, prominent scientists helped to justify the need for a GG study.

Kramer and Gudakunst organized a private meeting in June 1944 to discuss the possibility of a renewed effort to test GG for polio.[33] They invited the recently appointed Michigan State health commissioner and former medical director of the ARC, Dr. William DeKleine, to join the meeting.[34] DeKleine was a strategic choice, as he was knowledgeable about blood fractions and believed in the potential of GG.[35] He agreed to extend state resources to the proposed study, while Gudakunst pledged to negotiate a supply of GG and manage the "executive aspects of the study."[36] In turn, Kramer accepted the role of chief investigator.[37] DeKleine, Gudakunst, and Kramer emerged from the meeting with a plan to realize a GG study.

Gudakunst faced challenges negotiating a supply of GG.[38] He realized that convincing the ARC and NRC would be difficult, considering their prior rejection. His fears were well founded, as ARC medical director Dr. Foard McGinnes confided that he was not terribly enthusiastic about the proposal; however, McGinnes "promised" to release some GG as long as the NFIP could "present to him a program which [was] sound."[39] Appreciating that a rigorous clinical trial protocol was necessary, Gudakunst enlisted prominent researchers and statisticians Dr. Thomas Francis Jr., from the University of Michigan, and Dr. Kenneth F. Maxcy, from Johns Hopkins University. Gudakunst hoped that the ARC would honor its promise if he showed that prominent researchers were on board and the protocol was viable.

Gudakunst kept Basil O'Connor informed of ongoing developments. However, O'Connor's role at the NFIP was complicated by 1944, since he had agreed, at Franklin D. Roosevelt's request, to serve as chairperson of the ARC.[40] The dual appointment meant that O'Connor was well positioned to influence the release of GG, but he needed to appear beyond reproach. In his first report to O'Connor on July 13, Gudakunst set out a glowing history of GG research, explaining that scientists such as Stokes and Kramer were "quietly" conducting animal studies with encouraging results. Although he acknowledged the proposed study would be "fairly expensive," it was merited.[41]

Kramer's protocol was scrutinized and refined by a special committee assembled at NFIP headquarters.[42] Prominent polio researchers and NFIP grantees were present, including Drs. Maxcy and Francis, as well as Yale University polio researcher Dr. John R. Paul. The proposed clinical trial appeared to be a sizable venture, requiring two statisticians, six epidemiologists, and a range of supporting agencies.[43] Statisticians determined that a cohort of twenty thousand children was necessary to assess the blood fraction.[44] Half the subjects would be injected with 20 cc of GG, while the other half would be used as observed controls for comparison.[45] Kramer was optimistic, seeing no "difficulty in securing an adequate number of individuals."[46] However, some scientists took aim at the potential problems, reasoning that the trial would be "most difficult" to conduct and not yield "an unequivocal answer" on GG.[47] Despite concerns about data quality and logistical challenges, most committee members endorsed the study.

Kramer and DeKleine were convinced that the protocol would impress the ARC, but Gudakunst was less confident.[48] The potential problems raised by some committee members made him anxious. He wrote to O'Connor, warning that a sufficient supply of GG would be difficult to acquire, that the study could become "impractical," and that "there may be too many severe reactions" among the child subjects.[49] He also fretted about the possible shortage of nurses and epidemiologists to monitor suspected polio cases. To balance his misgivings, he reasoned that there was no known evidence of adverse reactions and that clerical staff could be used to substitute for nurses. For Gudakunst, the experiment had problems, but it was worthy of support.

Despite lobbying efforts, Kramer's GG experiment was rejected.[50] According to an unofficial report, the ARC committee turned down the study "because of a lack of nurses to do the necessary follow-up work."[51] To ascertain the validity of this rationale, Gudakunst consulted health professionals.[52] He learned that for every one thousand children injected with GG, approximately twenty nurses would be needed to conduct the follow-up work. As the Detroit health commissioner explained, "It would be impossible . . . to secure nurses for such an immunization program."[53] The commissioner reasoned that during a polio epidemic, nurses were needed on acute wards and not on an epidemiological study.[54] The official ARC decision arrived on August 21, 1944, and notably absent from this version was reference to insufficient nursing personnel. Instead, ARC committee members questioned the necessity of the study and its epidemiological value. Since most researchers at the time believed polio was a nerve disease that did not spread through the bloodstream, ARC committee members were unconvinced that injected antibodies could prevent the illness. They reasoned that the study was "not practical or justified by our present knowledge" and "could not be organized in time." Only studies "on animals including chimpanzees" would be supported.[55] By authorizing the release of a small amount of GG for animal experiments, ARC committee members were asserting that a human trial was premature and that basic research was more desirable.

Disappointed by the ARC committee decision, DeKleine sought to exploit Basil O'Connor's dual appointment. He wrote to O'Connor, outlining the proposal process and the involvement of the NFIP and requesting assistance in appealing the decision. DeKleine reminded O'Connor that many scientists supported a GG study for polio and that nurses could be made available.[56] He concluded by stating that he would "appreciate any effort which [O'Connor] as chairman of the Red Cross [could do to make] globulin available to this study."[57] As former medical director of the ARC, DeKleine knew its inner workings and skillfully situated his tabled experiment as a merited enterprise held up by red tape; he believed it could be swiftly extricated from the bureaucratic quagmire if only sound judgment and sufficient influence were applied. By circumventing Gudakunst and seeking O'Connor's involvement, DeKleine hoped that the ARC committee decision would be overturned.

While O'Connor considered his response, a severe polio epidemic swept through Chicago in late August 1944.[58] DeKleine used the occasion to urge O'Connor via telegram that an "opportunity for a field study" existed. O'Connor chose to uphold the official ARC committee decision and tempered the blow by answering DeKleine's letter in the role of NFIP president and his telegram as chairperson of the ARC. O'Connor explained to DeKleine that he was already acquainted with Kramer's protocol; however, he reasoned that "every consideration was given to the National Foundation's request and I am, therefore, compelled to abide by the decision of the American Red Cross."[59] Since the ARC was influenced and subsidized by the federal government during World War II, it needed to prioritize military over civilian initiatives.[60] A large study of GG would need to be shelved until the political, economic, and scientific climate changed.

Failure to receive ARC and NRC support led some researchers and public health officers to undertake private studies on institutionalized populations. During the winter of 1944, DeKleine and Kramer obtained a small supply of GG through Dr. Edwin Cohn's Harvard laboratory to conduct an experiment on children "identified as non-immune" to polio.[61] Gudakunst was made aware of their intentions and endorsed the overall plan. Unlike his earlier protocol, Kramer's small study pursued basic research questions in attempting to ascertain the duration of "neutralizing antibody" in blood and spinal fluid.[62] Kramer and DeKleine continued these "hush-hush" experiments into 1945, but failed to establish "definite proof" of efficacy.[63] Until the value of GG for polio could be established, researchers continued to experiment.

Like their counterparts in Michigan, Illinois public health officers undertook human GG studies. When a severe outbreak erupted in the summer of 1945 at Saint Vincent's Home for Children, a Catholic orphanage in Freeport, Illinois, the Department of Health responded swiftly.[64] The chief of the department's division of communicable diseases, Dr. Jerome J. Sievers, invited leading scientists to help control the epidemic using GG.[65] With degrees in medicine and public health, Sievers was experienced with polio outbreaks and had served the health department for many years in the capacity of district superintendent.[66] On August 25, 1945, he convened a special meeting at Chicago's Palmer House Hotel with ARC medical director Dr. Foard McGinnes and University of Michigan researcher Dr. Thomas Francis Jr.[67] Although sources do not reveal why NFIP representatives were excluded from the proceedings, it is likely that Sievers did not consider their involvement necessary.

At the meeting, McGinnes agreed to supply GG provided that Sievers permit the ARC to undertake a parallel epidemiological study.[68] Committee members also decided that the Saint Vincent's study would be conducted in secret, with "no publicity given." The orphanage's population of 427 residents, ranging in age from "nursery to high school," housed in a closed rural environment with a known history of polio outbreaks, was an ideal setting for a private experiment. As wards of the orphanage, residents were obliged to participate. With the assistance of Francis, the committee agreed to undertake a controlled study. Saint Vincent's children would be divided "as to age, sex, and other characteristics," injected alternatively with either saline or GG solution, and then injected two weeks later following the same method. Subjects would be closely monitored over the course of one month and required to provide stool and blood specimens.[69] Following unanimous committee approval, the ARC released the blood fraction for immediate application.

Since it was unknown when the polio epidemic at Saint Vincent's would peak, a field team headed by Sievers was assembled and equipped for deployment. On August 26, 1945, the team traveled the 114 miles from Chicago to Freeport. Upon arrival at Saint Vincent's, the scientists evaluated the institution and explained the "experimental basis" of the study to the prefect, Father Kennedy, and the Mother Superior. Father Kennedy was not terribly enthusiastic and believed "the proposed inoculations were not necessary," since the epidemic appeared to be waning. State health officers disagreed and insisted that

Father Kennedy "abide" by their recommendations.[70] Children six years of age and under were randomly assigned an injection of 10 cc of either saline or GG serum, while those over six years received 20 cc. Some restrictions were placed on the experiment when the prefects objected to the taking of blood samples.[71] After the injections were completed, a follow-up team remained at the site to document complications and emergent cases of polio.[72]

The results of the Freeport experiment were inconclusive. The administration of GG had not reduced the incidence of polio, and according to Francis, the experiment had not provided "significant information as to the effect [of GG] upon the disease." The absence of blood specimens had a deleterious effect on the evaluation, since antibody titration tests could not be undertaken. Despite emphasis on statistical rigor, the cohort was not of sufficient size to yield meaningful conclusions. With only stool samples to analyze, Francis conceded that "we didn't learn anything except the fact that you could give gamma globulin."[73] Beyond scientific factors, the Saint Vincent's experiment revealed opposing concerns and priorities. While Saint Vincent's prefects were willing to cooperate with researchers, they also saw science as negotiated territory when it left the confines of the laboratory. They did not allow the scientists to carry out the full range of intended evaluations, but they did allow the use of injected controls. The lesson to scientists was not lost: if they sought to conduct future studies, they would need to anticipate the anxieties of parents or guardians. Since the experiment did not return conclusive results, uncertainty about the efficacy of GG in preventing polio paralysis remained.

Following the Freeport study, few medical researchers attempted to assess GG for polio. The blood fraction was not available in sufficient quantities, and the resources required to manage a large experiment were difficult to assemble at a time of war and recovery.[74] The landscape of biomedical funding was also changing. The National Institutes of Health expanded its grant system, favoring research into cancer, psychiatric health, and heart disease.[75] Moreover, the death of Gudakunst from a heart attack in 1946 left the NFIP in transition, while incoming research director Dr. Harry Weaver assessed past and future priorities.[76] Formerly an anatomy instructor at Wayne State University, Michigan, Weaver was a visionary who sought to position the NFIP as the leader in polio research initiatives.[77] The history of disappointment and expense relegated GG to a low institutional priority.

Without conclusive data on GG for polio, American clinicians were left to make their own decisions about whether to use GG during an emergency. When Houston, Texas, was visited by a severe outbreak in June 1948, pediatricians, demoralized by their limited means to halt the epidemic, administered GG to family contacts of polio patients in the belief that it might guard against the virus.[78] Attending the epidemic on behalf of the NFIP was consultant Dr. William McDowell Hammon, who became intrigued by the potential of GG.[79] Although there "were no controls and no valid conclusions" were reached, the outcome reportedly followed the "pattern which [was] to be expected should the antibody have a protective effect."[80] As an ad hoc intervention, GG provided some desperate families with hope. However, for a few resolute polio researchers, an ardent faith in blood fractions kept the dream of a large GG clinical trial alive. One problem remained: Who would lead such an audacious experiment?

The Apprenticeship of William McDowell Hammon

A witness to the 1948 administration of gamma globulin in Houston, Texas, Dr. William McDowell Hammon emerged as the leading contender to spearhead a pioneering medical experiment. Born in 1904 in Columbus, Ohio, Hammon spent his formative years in Conneautville, Pennsylvania.[81] His father, an engineer who taught religion at theological institutions, encouraged his son to enter the Methodist ministry.[82] After high school and in accord with his parent's wishes, Hammon embarked on an ecclesiastical career and was ordained a Methodist minister.[83] At the age of twenty-two, inspired by religious convictions, he applied to work as a missionary in the Belgian Congo.[84]

At the time, the Belgian government ruled the Congo under an authoritarian regime that restricted Africans from an active role in legislation or governance. Resource exploitation was rampant, as private American and European companies invested in the region and established large plantations, mining operations, and livestock farms. Many Congolese lived in poverty and worked as indentured laborers; rebellions were not uncommon, as workers attempted to resist abuse and exploitation. Malnutrition and disease were rife. The Methodist mission in the Congo attempted to fill a void in the state apparatus by providing basic education and medical outreach.[85]

Before embarking on his higher calling, Hammon was enrolled in the School of Tropical Medicine in Antwerp, Belgium, for training in what one contemporary referred to as "the medical end of the missionary work."[86] After completing a rudimentary course, Hammon was awarded a "license to practice tropical medicine and perform minor surgery" and sent to a Congo mission to direct a medical dispensary.[87] Echoing themes introduced in Joseph Conrad's *Heart of Darkness*, Hammon "became a printer, blacksmith, carpenter and cook" and recalled the challenges of ministering to "the physical and spiritual ills of his converts." As he explained, "It's a helpless feeling to be 400 miles from the nearest doctor with dying people looking to you—who didn't know enough—for help."[88] After nearly five years in the Congo, he became disaffected by his rudimentary medical training.[89] According to Hammon, his missionary experience incited in him a quest to "become a doctor—to prevent rather than cure."[90] Missionary zeal combined with the demands placed on a foreign medical worker pressed Hammon to further his education.

In 1930, Hammon returned to the United States. He enrolled in premedical studies at Allegheny College in Meadville, Pennsylvania, and was subsequently accepted to Harvard University Medical School. He completed his program in 1936 and moved to Pittsburgh to begin an internship at the Allegheny General Hospital. However, the Great Depression posed challenges for medical students, since internships were unpaid and the fall in personal incomes limited the number of job opportunities.[91] It was at this juncture that Hammon's former Harvard professor, the noted microbiologist Dr. Hans Zinsser, invited him to return for graduate studies and to serve as his teaching assistant.[92] Hammon embraced this opportunity and resumed his studies, earning a master's and doctor of public health in 1938 and 1939, respectively.[93] Zinsser's intervention permitted Hammon to begin a career in medical research.

Although Hammon intended to return to the Congo, his plans were dashed by the outbreak of World War II in Europe.[94] Once again, through Zinsser's aid, Hammon was invited to remain at Harvard and become an instructor in epidemiology. As a budding scientist, Hammon became a determined and "stubbornly patient worker," and his laboratory research yielded notable discoveries.[95] He conducted research into staphylococcal toxins and collaborated with eminent scientist Dr. John F. Enders on the first feline distemper vaccine.[96] As the nation prepared for the prospect of war, Hammon enrolled in the U.S. Army medical commission system and became a consultant to Surgeon General Dr. Thomas Parran Jr. on the Virus and Rickettsial Diseases Commission.[97] Following nearly a decade of training, Hammon had demonstrated his expertise in laboratory and field research.

In the spring of 1940, Dr. Karl F. Meyer, director of the University of California's Hooper Foundation medical research unit and a close friend of Zinsser, offered Hammon a position as an assistant professor of epidemiology. Meyer reportedly sought Hammon for professional and personal reasons, since he wanted to replace Dr. Beatrice F. Howitt, a woman scientist he deemed too "independent," with a genial and competent male researcher experienced with viruses of the central nervous system.[98] The gender biases that pervaded the scientific community provided Hammon with a rare professional opportunity.[99] As one of Hammon's former students remembered, "Dr. Meyer hired [Hammon] and said, 'Okay, pack up your family.'"[100] In July, the Hammons set out from

Pennsylvania for California; however, their journey was altered when Hammon was asked by Meyer to send his family on alone and divert his course to Washington State to assess an encephalitis epidemic in Yakima Valley.[101] Channeling his former missionary energies into his new position, Hammon's professional ambition and commitment were evident well before his arrival at Berkeley.

Hammon enjoyed the University of California and its opportunities for career advancement. Influenced by his Yakima Valley investigation, Hammon's research at the Hooper Foundation focused on encephalitis, but soon broadened into other viral diseases, including poliomyelitis.[102] Although he privately considered polio of minor significance, because of its low incidence rate compared with other childhood diseases, his work soon attracted the attention of the NFIP. According to Hammon's former student, "The National Foundation was interested because polio and encephalitis still were confused both diagnostically and epidemiologically."[103] Since research grants were notoriously difficult to obtain in the 1940s, the willingness of the NFIP to generously fund polio research probably influenced Hammon's research agenda.

By 1945, Hammon became more involved with teaching and was promoted to associate professor of epidemiology at the School of Public Health.[104] His interest in polio increased, and he communicated with leading virologists

Figure 4 Dr. William McDowell Hammon, 1951. Courtesy of the March of Dimes Foundation.

about causation theories.[105] He and researcher Dr. Thomas Francis Jr. cultivated a friendship during these years, which served as a forum for ideas and debate. In gratitude for Francis's hospitality and laboratory tour during an April 1945 visit to Ann Arbor, Michigan, Hammon wrote cheerfully: "The two days with you were most pleasant, and I shall long remember the fine entertainment which you provided. I shall remember in particular the evening we went 'slumming' and although we did not solve the question of the epidemiology of poliomyelitis, we did make progress with frog legs and fried chicken."[106]

In addition to his academic responsibilities, Hammon's 1946 appointment to the U.S. Army Tropical Diseases Commission led him on numerous expeditions to the Pacific, including to Japan, China, Korea, and Guam.[107] During these forays, Hammon tracked viral epidemics and theorized about methods to control polio. That same year, he was recognized for his civilian service in wartime when Harry S. Truman awarded him the Presidential Medal of Freedom.[108] Hammon continued to publish; he authored several medical textbook chapters and served as associate editor of the *Journal of Immunology*. After nine years at Berkeley, Hammon had amassed considerable authority as dean of the School of Public Health and assistant director of the Hooper Foundation.

Although established in California, Hammon seized on a unique opportunity offered in Pennsylvania. In 1948, Dr. Thomas Parran Jr. retired as surgeon general and became dean of the University of Pittsburgh's newly developed Graduate School of Public Health.[109] As part of his ambitious vision, Parran invited recognized researchers and instructors to help him develop the public health program and transform it into a center of excellence.[110] Since Parran knew Hammon through mutual colleagues and prior consulting activities, he was keen to enlist him to head the Department of Epidemiology and Microbiology.[111] Hammon accepted the appointment in 1949, perhaps reasoning that it would offer new opportunities and reconnect him with the northeast scientific elite. Pennsylvania was also his former home state.[112] However, when Hammon departed, he left a considerable leadership and funding void.[113] Despite the consequences for Berkeley, Hammon was positioned in Pittsburgh to shape the curriculum and a new generation of public health students. It was at this time of transition that he began to imagine a GG experiment for paralytic polio.

Imagining a Gamma Globulin Field Trial

Hammon's interest in GG for polio was not inspired by his own laboratory investigations, but by a series of chance encounters and professional assignments.[114] While attending a polio outbreak on behalf of the NFIP in the late

1940s, Hammon was asked by one physician, "Why isn't gamma globulin used to treat the disease in its preparalytic stage? Why isn't it used in the prevention of poliomyelitis?" Although Hammon knew that GG was used for the prevention of hepatitis and measles, he reportedly explained to the physician that it "had not been shown to be effective [against polio], and its use was therefore not recommended at present." Although he later "worried" about this conservative answer, a subsequent occasion strengthened his commitment to a GG study.[115]

The incident came on November 2, 1949, when Hammon was asked by the American Academy of Pediatrics to prepare and deliver a paper at its symposium two weeks hence.[116] Since polio was a prevailing concern, and as Hammon was considered to be an expert on viruses, organizers persuaded him to deliver a paper with the tentative title "Possibilities of Specific Prevention and Treatment of Poliomyelitis."[117] Hammon was concerned by the assigned topic and acknowledged to University of Cincinnati polio researcher Dr. Albert B. Sabin that it was a subject he "would like to have dodged." Hammon understood that the title was presumptuous, and he feared that speculation might get him "in trouble" with other researchers. He begged Sabin for "any kind of preliminary note" or "anything in press" that he could incorporate into his symposium paper.[118]

Sabin shared some of his research with Hammon, but was reluctant to provide citable material. Sabin noted that a compound used in the treatment of malaria showed promise in preventing poliomyelitis in monkeys, but that he had "not published anything on it until [he] could be more certain of the effect." Although Sabin speculated that this finding might lead to new compounds for polio prevention, he asked Hammon not to "mention this work in [his] formal presentation." Hammon was left alone to contemplate the current state of polio research and fume over the topic he had agreed to present to pediatricians.[119]

The impending deadline forced Hammon to consult medical and scientific journals. During his literature review, he reasoned that GG for polio had been overlooked and that a field trial appeared long overdue. "Although for years, I had been working in the laboratory and the field on immunologic and serologic responses to poliomyelitis infection," he later recounted, "the more thorough review of the literature and the careful thought required to cover adequately the assigned subject, suddenly led me to wonder why passive serum prophylaxis in poliomyelitis had not been given more serious consideration."[120] Recent scientific developments appeared to justify the evaluation of GG. Dr. John F. Enders and his colleagues at the Children's Hospital in Boston learned that poliovirus could be cultured in non-nervous tissue, proving that polio was a systemic disease that might be prevented through immunization.[121] Dr. David Bodian at Johns Hopkins University discovered that the three known types of polio could be neutralized with GG.[122] What remained unknown was the role that passively conveyed antibodies played in the pathology of polio, and the duration of such immunity.[123] In Hammon's symposium paper, later published in the journal *Pediatrics*, he attempted to build support for a clinical trial.[124] Despite his best efforts, Hammon observed "no stampede by fellow scientists to perform such an experiment."[125] The combined hesitancy of polio researchers, increasing incidence of polio, and unresolved questions over GG efficacy in humans led Hammon to imagine undertaking an experiment himself. Interactions with clinicians and a critical assessment of literature had seeded his fascination with GG.

Hammon rationalized the use GG for polio on his theory of depreciating immunity. Many researchers at the time reasoned that the immune response to polio was similar to that of mumps, chicken pox, or measles, in which a survivor developed antibodies and lifelong protection. Studies by Yale University researcher Dr. John R. Paul among Alaskan Inuit communities in 1949 appeared to uphold the theory of durable polio immunity.[126] However, Hammon disregarded this model as an assumption; instead, he postulated that polio immunity followed that of streptococcal infections or diphtheria, in which immunological resistance depreciated over time and was eventually lost.[127] Although active immunity (a form of resistance in which the body produces its own antibodies to a disease) was associated with surviving an infection, he believed it did not guarantee lifetime protection.

Hammon's theory of depreciating polio immunity was grounded on observations made while consulting for the U.S. Tropical Diseases Commission.[128] In November 1948, a severe polio epidemic swept through the small Pacific island of Guam, affecting a disproportionately high number of American personnel and their families, but few indigenous people. The last recorded polio outbreak on the island had occurred in 1899, following the Spanish-American War.[129] The dearth of paralytic polio among Guamanians did not come as a surprise to Dr. C. K. Youngkin, the director of the Department of Public Health in Guam, who reportedly had "never been able to find any evidence of paralytic poliomyelitis among Guamanians although he . . . looked for it carefully."[130] Intrigued by the apparent immunological disparity, Hammon and Youngkin conducted studies among the indigenous population and found that most harbored antibodies to poliovirus. Guamanian children also developed antibodies at a much younger age than children on the American mainland. Since Hammon considered Guam a type of Darwinian Galapagos because of its "limited, semi-isolated population," he believed it was the perfect locale to yield truths about polio.

Based on his studies in Guam, Hammon reasoned that if polio immunity was truly durable then the disease should have died out on the island. Once the virus was extinct, the local population would lose protective antibodies and be vulnerable. However, since the poliovirus was endemic in Guam, it seemed plausible to Hammon that this was evidence of constant reinfection and a carrier state associated with the depreciating immunity.[131] His explanation for why the indigenous population acquired immunity early in life and maintained it without complications was that it was because of their isolation and living conditions. Since mainland Americans did not have these mitigating factors, Hammon believed they were perennially vulnerable.[132] Not only was his theory based on a seductive blend of comparative epidemiology and socioeconomics, it was plagued by assumptions. His contention that Guam's indigenous population was "semi-isolated" was tenuous, as the island was an active way station and former Spanish colony since the 1660s.[133] Moreover, his use of bacterial models of disease instead of a viral disease model such as measles was based on speculation and armchair observations.[134] The presumed uniqueness of Guam and its local population was the foundation of Hammon's novel theory.

Hammon also justified clinical experimentation with GG on the possibility of producing passive-active polio immunity.[135] This model of disease prevention was loosely based on the earlier infectious hepatitis studies undertaken by Dr. Joseph Stokes Jr., who imagined that GG injections could create a threshold of immunity sufficient to prevent severe illness, but not infection.[136] It was hypothesized that the mild or asymptomatic illness would inspire the immune system to generate its own antibodies and confer active, but temporary, immunity. Although the theory of passive-active immunity was not proven, it fit within Hammon's model of the disease.

Hammon had little faith in brazen plans to develop an effective polio vaccine. He believed that any polio vaccine would be inherently dangerous and that the potential "risk of accident to thousands to protect a single individual" was unethical. Even if a vaccine was discovered, Hammon believed that GG would be a more expedient and safer intervention. Since he speculated that polio immunity declined over time, he reasoned that yearly vaccination for a comparatively rare affliction like polio would be wasteful. "If vaccination is practiced annually and by good fortune should give 100 per cent protection," Hammon mused, "this means that approximately 11,000 children would have to be immunized annually to prevent the annual average occurrence of each single fatal or permanently disabled child. . . . Is this practical or justifiable?"[137] For Hammon, GG was a better alternative, since it would be given only during an epidemic and "its effect would be immediate and would represent no danger to any child."[138] GG was not merely a stopgap solution on the road to a vaccine, but the only solution to paralytic polio. Although not all scientists agreed with Hammon's theories or assumptions, his conviction inspired others.

Hammon rationalized a clinical trial with GG on the basis of public health pragmatism.[139] Since GG was available for clinical use and was shown to prevent other diseases, it appeared to be a natural choice for a researcher concerned with expediency. Furthermore, Hammon appears to have approached the problem of polio control in the mindset of a public health officer, rather than a laboratory scientist. His former role as dean of the School of Public Health at Berkeley had moved him further away from the laboratory and into administration and teaching. Evidence suggests he was less interested in assessing immunological mechanisms and more interested in applied research; he was not concerned with the prevention of polio infection, but rather the reduction of its most egregious paralytic effects.[140] His preference for applied research showed a pragmatic approach to a viral threat.

Informed by personal observations, literature reviews, and public health pragmatism, Hammon was committed to assessing GG. He developed five experimental protocols, derived from his theories, ranging from GG nasal sprays to various injection plans. Concerned whether the proposed studies were "morally and legally tenable," Hammon rated them in order of preference, with the less desirable options "put into effect in a foreign country where the level of public and professional knowledge is much below that of the United States."[141] Like other American researchers who planned to conduct experiments in foreign regions, Hammon was anxious about how his proposed studies would be perceived by others and whether they might cause adverse health reactions.[142] Although Hammon showed awareness of ethical imperatives, he did not see all peoples as deserving of the same standard.

After fielding his proposed protocols at a special private meeting on gamma globulin, Hammon settled on a human experiment to assess whether the "largest" practical injected dose of GG protected children from polio paralysis.[143] By the 1950s, the concept of a randomized double-blind placebo-controlled trial was gaining support. The methodology, pioneered by British statistician Ronald Aylmer Fisher in *The Design of Experiments* and Dr. Austin Bradford Hill in the *Principles of Medical Statistics*, led Hammon to plan the random assignment of either GG or placebo.[144] The study would be undertaken in the midst of an epidemic, and thousands of subjects would be enrolled to achieve statistical significance. By comparing polio incidence in the two groups, Hammon believed the efficacy and duration of immunity conveyed by GG could be ascertained.[145] Although he was uncertain whether his protocol was legally justified, he believed it was practical and ready for consideration.[146]

Chapter 2

Building Consent for a Clinical Trial

"The thing I am getting to is if you start out and do this in an open population all hell is going to break loose and you know that just as well as I do," Dr. Joseph E. Smadel, chief of the Department of Virus and Rickettsial Diseases with the U.S. Army Service Graduate School, cautioned Dr. William McD. Hammon in February 1950.[1] Although Hammon had supporters, he also faced criticism and setbacks that would require the support of the NFIP to ensure his study could be brought to an open population. Divisions in the scientific community and evidence of health risks, as well as serious methodological and logistical problems, threatened to derail the project. After several prior disappointments, NFIP officials moved carefully and worked with Hammon to build scientific approval and defuse criticism. During this period of debate and negotiation, the NFIP helped Hammon realize his ambition. How did medical researchers respond to Hammon's proposed GG field trial? How did he and NFIP officials respond to the concerns raised by scientists? This chapter examines the discussions and how consent for the study was eventually constructed through compromise and coercion.[2]

Powerful Interests and American Realities

Unlike the gamma globulin field trials proposed in 1943 and 1944, enthusiasm for undertaking a GG study in 1951 was fueled by personal and institutional agendas and facilitated by societal realities. At a time when the incidence of polio was increasing and virologists were slowly unlocking the secrets of the disease, Hammon was eager to offer his contributions to society and science.[3] He was confident in the blood fraction, since it was shown to be safe and effective at fighting measles and hepatitis.[4] As the recently appointed head of the Department of Epidemiology and Microbiology at the University of Pittsburgh, Hammon also wanted to prove himself as a worthy leader. The favorable media attention, generous funding, and scientific publications resulting from a large clinical trial would not only bring the university prestige, but also uphold the wisdom of Hammon's appointment. His boss, Dean Thomas Parran Jr., was building a public health empire at Pittsburgh, and any glory that could be harnessed for his institution was sure to strengthen Hammon's position.[5]

Tensions with a prominent Pittsburgh colleague also encouraged Hammon to be decisive about his future. When he began his appointment on February 1, 1950, his presence was not universally welcomed.[6] Dr. Jonas Salk, who had arrived at the university in 1947 as director of the Virus Research Laboratory, was angered by Parran's decision to hire Hammon. As Salk remembered: "I had come from a school of public health and a department of epidemiology, it seemed to me that my background and experience brought me within reason as a candidate for the job. . . . I remember being quite upset at not having been considered."[7] Hammon's more senior appointment and parallel interest in polio instigated tensions with Salk, which were exacerbated by their differing theories about prevention.[8]

The rivalry with Salk became sufficiently tense that it created an atmosphere of antagonism that threatened Hammon's leadership. "There was no love lost between Bill Hammon and Jonas Salk," remembered Salk's research collaborator Dr. Julius Youngner. "Needless to say, this did little to create an atmosphere of collegiality at the University."[9] Annoyed by Hammon's belief that a polio vaccine would be dangerous and impractical, Salk was resentful of Hammon's consultations with Virus Research Laboratory staff.[10] Salk presented his frustrations to NFIP research director Dr. Harry Weaver in the hopes of intervention. "You will recall my talking to you," Salk reminded Weaver, "specifically about the possibility of conflict with Hammon." Salk reconstructed one wearisome incident: "Hammon inquired of one of the people on our staff about the various instruments that we use for autopsying monkeys, and various other technical questions. At that time he also inquired about 'stomach tubes' or catheters that might be used in monkeys."[11] Although Weaver assuaged Salk's anger, Hammon remained unwelcome at the Virus Research Laboratory.[12] The lingering discontent between the researchers probably inspired Hammon to assert his prominence in the field of polio research by leading a medical study.

The lobbying efforts of GG proponent Dr. Joseph Stokes Jr. also encouraged Hammon to be decisive.[13] During the late 1940s, Stokes petitioned the NFIP to sponsor a GG study for polio. Although Weaver was sympathetic to the idea, not all NFIP officials were enthusiastic. As Stokes explained to Hammon, "They have been listening with more interest, but still not very sympathetically."[14] Since no medical researcher emerged to lead a GG study, Stokes privately encouraged Hammon to step forward and submit a research proposal to the NFIP.[15] To cement his convictions, Stokes offered to collaborate on the study if it received institutional backing.

Like Stokes, some scientists reasoned that a field trial was necessary because of the unchecked clinical use of GG for polio. The availability of the blood fraction, combined with evidence that it protected against measles and hepatitis, led some doctors to offer it to household contacts of polio patients, hoping it might guard against the virus. Hammon's optimistic article in *Pediatrics* in 1950 unintentionally justified such ad hoc clinical use of GG.[16] Indeed, in September 1950, Stokes was notified by a North Carolina pediatric consultant about enterprising doctors billing desperate parents for GG injections. "Some of the physicians are buying it and charging fairly stiff fees for its administration," the consultant reported. "Parents are demanding that something be done for the children who have been exposed and one pediatrician in the Durham area gives it routinely."[17] Stokes advised the consultant to dissuade doctors from using GG for polio until more was known.[18] For such researchers, only a rigorous clinical trial could end uncertainty and false hope.

Other researchers were worried that the unchecked use of GG would lead to speculative publications based on dubious results. University of Toronto's Connaught Laboratories researcher Dr. Andrew J. Rhodes warned, "There will be all sorts of half-baked papers coming out in the next five years by pediatricians who have not the first conception of mathematics. . . . I think we ought to try and give them some guidance."[19] His opinion was linked to a mid-century belief that scientists evaluated and created knowledge, while clinicians implemented proven methods.[20] For Rhodes, the data set generated by a field trial would help ground clinical practice and uphold the established role of medical research.

Hammon's proposed GG study also found support among researchers disquieted by reports of rushed and hazardous human polio studies. At the Round Table Conference on Immunization against Polio in March 1951, Dr. Hilary Koprowski, a polio researcher based at Lederle Laboratories, presented his findings on the testing of a prototype oral attenuated live-virus vaccine on children. "I made my presentation immediately following lunch," Koprowski remembered. "[Dr. Thomas Francis Jr.], who was having a postprandial nap, looked up at my slides showing vaccination of children, and asked [Dr. Jonas] Salk (who was sitting beside him), 'What monkeys are these?' 'These are not monkeys, but children,' Salk answered. Francis woke up immediately and snoozed no further."[21] Most attendees were shocked at the audacity of Koprowski's experiment and its ethical implications.[22] Koprowski recalled: "Sabin asked me, 'Why did you do this? It may be a dangerous approach since there is a Society for the Prevention of Cruelty to Children.'" Stokes was likewise incredulous and inquired as to whether Koprowski was "afraid that the Society for the Prevention of Cruelty to Children would 'go after'" him.[23] The ensuing criticism of Koprowski showed concern over the conduct of pediatric experimentation and roused anxiety in many attendees. Such concern was prescient, since Salk would later test his prototype killed virus vaccine on children at the Polk State School and D. T. Watson Home for Crippled Children.[24] In contrast to Koprowski's audacious human studies, the modest scheme put forward by Hammon appeared cautious. During the meeting, Hammon outlined his protocol, whereby during "a large poliomyelitis outbreak, volunteers would be quickly injected with a predetermined dose of gamma globulin" or placebo. Delegates discussed the proposal and it "was considered by a majority present as suitable in its general concept."[25] Although Salk and Sabin also reported results of their laboratory research, it was apparent that an effective polio vaccine was years away. With evidence of risky human experimentation on one hand and measured laboratory research on the other, Hammon's proposal held the middle ground; his proposal was expedient and took advantage of a substance already in clinical use.

Despite initial reserve, NFIP officials were growing increasingly open to supporting a GG study in 1951. With no immediate promise of a vaccine and rising medical treatment costs ranging from $900 to $2,000 per paralytic patient, O'Connor was keen to take control of financial costs.[26] Although contributions to the March of Dimes had increased over the years, the $30 million raised in the 1950 campaign did little to cover polio treatment expenses.[27] Moreover, the NFIP was desperate to showcase progress in medical research. For nearly a decade, NFIP research committees allocated millions of publicly donated funds to scientists with few momentous breakthroughs.[28] The hope and excitement surrounding a GG study promised to help the charity justify a costly research program to its donors.

NFIP personnel were also confident in Hammon as the chief scientific investigator. From the early 1940s they had awarded him research grants when he was based at the University of California. By the late 1940s, Hammon was working for the NFIP as an epidemiological consultant, offering advice to communities in the throes of polio epidemics.[29] O'Connor and his staff trusted Hammon and knew that investing in him was not a dangerous gamble.

The relationship between the American Red Cross and NFIP also made a GG study increasingly possible. O'Connor's departure from the ARC, after serving as chairman (1944–1947) and president (1947–1949), assured that he could leverage his clout on behalf of the NFIP without appearing to unduly influence internal committees.[30] His knowledge of the ARC assured that requests received serious consideration. Furthermore, the ARC held a rare abundance of dried GG left over from World War II, which could be allocated to a medical study.[31] ARC vice president Dr. Foard McGinnes reasoned that "it would be much easier to set aside an amount for the Foundation . . . now than it will be later on."[32] Cooperation between the NFIP and ARC provided the institutional commitment necessary to realize a large study.

Scientific Critiques and Health Risks

Although Hammon had many supporters, not all medical researchers were enthusiastic about his proposed GG study. Between 1950 and 1951, criticisms were raised by prominent researchers at scientific gatherings, through correspondence, and at meetings of the NFIP Committee on Immunization. The committee proved to be an important peer review forum for Hammon's proposed study; it was originally inspired by Dr. Sam Gibson of the ARC and established by Dr. Harry Weaver in April 1951.[33] Although the committee did not enjoy de facto authority over NFIP research mandates, its de jure authority was remarkable.[34] Through such forums, medical researchers scrutinized the proposed GG study and revealed a number of troubling aspects requiring Hammon's attention.

Some scientists were troubled by Hammon's plan to conduct the experiment on an open population. Dr. Joseph E. Smadel raised concerns about the danger of sensationalist publicity and logistical challenges at a 1950 meeting on gamma globulin. He asked Dr. Albert Sabin whether it was possible to instead find a private institutional setting for the study; however, since thousands of children were needed, Sabin reported that there was no institution large enough. Smadel remained nervous. "I didn't speak out in favor of it," Sabin rejoined dryly.[35] Since a large clinical trial had never been attempted on an open population before, scientists like Smadel were uneasy about the public response and media scrutiny.

Other researchers were reluctant to support Hammon's study because his protocol was based on what children would tolerate, not on what might provide protection against paralysis. In 1950, medical scientists did not know the level of antibody (titer) needed to prevent polio paralysis, the duration of such protection, or how lower titers affected the virus.[36] At the meeting on gamma globulin Sabin asked pointedly, "How much of this stuff has to be introduced, is there a level of antibody in which you will do absolutely nothing at all?"[37] Animal studies suggested that 1 cc of GG was needed per pound of body weight, but a comparable volume for children would be hazardous. Dosages needed to be based on practical as well as public relations considerations.[38] As Hammon explained, "We have got to get them fast, we have got to get them in tens of thousands, and these larger amounts are absolutely impractical from the standpoint of administration in this type of experiment. If we are going in for 40 cc for a forty-pound child, I just throw up my hands and say that you and I cannot get tens of thousands of children within days to stand in line and get that kind of dose."[39] Hammon favored predefined dosages of 4 cc to 11 cc, where younger children received the minimum dose and older ones received the maximum.[40] "Heck, he couldn't have given much more without having a first-class revolt on his hands," recalled Rockefeller University researcher and NFIP adviser Dr. Thomas Rivers. "I don't think there is a kid in America who would have stood, let us say, for an inoculation of 10 cc in each buttock without complaint—it just would have been too damn painful, and in the end it would have made the trial impractical."[41] Although Hammon's dosage schedule was practical, it did not convince most scientists that the efficacy of GG for polio could be accurately determined.

With related concerns about dosages, some scientists were skeptical of Hammon's empirical approach, believing that his study would not answer basic scientific questions.[42] In particular, his study would not assess prior immunity or inapparent infections among participants, nor would it measure titer levels at different GG dosages or explore the underlying mechanism of passive immunity. The field trial would be applied research without the foundation of basic research. At one meeting of the Committee on Immunization, Johns Hopkins University researcher Dr. Howard Howe protested: "It . . . seems to me that the cart is ahead of the horse and is a very heavy one to boot. To study subclinical immunization is less pretentious and far easier to do."[43] For Howe, it was necessary to understand whether and how GG affected humans before it was administered in large quantities to healthy children. Sabin agreed, arguing that laboratory research was essential before a clinical trial was undertaken. "In planning such an experiment . . . we ought not to be working in the dark," he cautioned. "We ought to know to what extent the large dose will be diluted in the body, and we should plan to give a dose which will yield a level of antibody in the serum that is equivalent to the minimum found in the adult population."[44]

Most scientists advised Hammon to modify his protocol to explore basic research questions on polio immunity. Johns Hopkins University polio researcher Dr. David Bodian suggested that Hammon conduct neutralization tests with blood samples gathered from a small cohort of children to assess how GG affected antibody levels. Salk concurred, stating that such knowledge would offer valuable insights into human immunity. Hammon rejected the idea, claiming that attention to titer levels and inapparent polio infections would be a distraction. "If we concentrated on the inapparent infection and possible interference," he argued, it "would possibly not give us the answer to the other question. I would not want to run the experiment that way."[45]

The stalemate was resolved by Dr. Thomas Rivers. "Maybe we have been too scientific so far," he mused. "I fear that if we had waited for scientific data in regard to the value of measles convalescent serum we would still not be using it. Even now, I cannot figure out why convalescent serum works."[46] For Rivers, Hammon's proposed GG field trial, however flawed in its design, was a useful exercise that might accidentally yield important knowledge. "I got up and made a statement that on its face sounds idiotic," Rivers later reflected at an interview. "I had become convinced that we could talk about [proposals] for another ten years and still not reach a conclusion. Under the circumstances, I felt that the only way to reach a conclusion was to put the experiment on—I was probably influenced in making my statement by Sir Francis Bacon who is once reputed to have said . . . 'Don't think—experiment!'"[47] Although not all scientists agreed, Rivers's timely intervention opened up the possibility for a compromise.

Although evidence suggests Hammon did not initially consider the health risks associated with his experiment, he was soon forced to consider their implications. Improved disease reporting, epidemiological surveillance, and mid-century expansion of research laboratories provided data and resources to evaluate medical interventions and their effects on people.[48] Because of these changes, the potential health risks associated with the GG study were anticipated and deliberated among scientists.

The danger of crowd contagion was a serious concern for the NFIP, the Committee on Immunization, and scientific advisers. During summer months, when polio outbreaks were most prevalent, children were instructed to stay away from public places such as movie theaters or swimming pools.[49] Notices were printed in magazines and newspapers, such as the *New York Times*, advising people to "avoid crowds and places where close contact with other persons is likely."[50] Children's summertime activities became highly regulated in the belief that it might limit exposure to the poliovirus.[51] To educate parents about crowd contagion, the NFIP developed a series of captivating posters with bold line drawings and memorable slogans.[52] In addition to portraying concepts such as "Don't Get Tired," "Don't Get Chilled," and "Keep Clean," the posters advised viewers to "Avoid New Groups."[53] Many radio shows promoted these pointers and small-town newspapers devoted several column inches to printing these graphic designs.[54] Since Hammon's GG study would require healthy children to attend clinics at the height of a polio epidemic, the convergence of large crowds at a public venue would facilitate the spread of poliovirus.[55] Participation in the study would therefore flout medical advice and decades of public practice.

At one Committee on Immunization meeting, NFIP medical director Dr. Harry Weaver reminded Hammon that there was a "danger of bringing the crowd together and increasing the risk in that group."[56] Echoing this sentiment, one committee member warned that "I think there is dynamite in any way we move in this thing and dynamite in the nature of the reaction of the parents of the child that is used as a control; should that child develop poliomyelitis and if the parent is a psychopath he might get a shotgun and go looking for somebody."[57] Unperturbed, Hammon acknowledged that he actually expected to see an increased incidence of polio among study participants; the regrettable risk of crowd contagion was the price of acquiring data to fight polio.

Members of the committee offered suggestions to help Hammon reduce the need for congested public clinics. One member recommended that children be selected by a door-to-door canvassing campaign from which only those residing in odd numbered houses would be invited to participate.[58] Although this approach reduced the chance of crowding, Hammon imagined serious public relations problems. "There will be political pressure," he retorted. "All the influential citizens that come in that group that want their children to get it and could not would go to the mayor, the councilmen and so forth, and all kinds of pressure would be exerted to see that they get it."[59] For Hammon, only a double-blind placebo-controlled trial could provide a fair and defendable method. In failing to reach a compromise, some committee members advised against using a placebo-controlled study. Sabin proposed that a fixed number of children be injected with GG on a "first come, first served" basis, while the remaining children became observed controls. Hammon pointed out the statistical bias: "The persons at the least risk," he explained, "will be the least likely to come

Figure 5 National Foundation for Infantile Paralysis Polio Pointers, 1951. Courtesy of the March of Dimes Foundation.

to the clinic," thereby skewing the data set.[60] After hours of debate, researchers could not agree on a strategy to reduce the risk of crowd contagion and left the matter to Hammon to ponder.

Evidence that certain pediatric injections could cause paralytic polio also concerned researchers.[61] Among the most influential investigators of polio provocation was Dr. Gaylord W. Anderson, who held a PhD in public health from Harvard University and was director of University of Minnesota's School of Public Health. Anderson was a specialist in communicable diseases with wartime experience in medical intelligence.[62] Applying this expertise to polio provocation, Anderson examined 2,709 case histories from a 1946 polio outbreak; based on family interviews, Anderson determined that "in poliomyelitis patients who have received some antigen [injection] during the month prior to onset there is a high degree correlation between site of paralysis and site of injection."[63]

The mechanism behind polio provocation was debated, and as an editorial in the *Journal of Pediatrics* conceded, "just how this is brought about remains a mystery."[64] According to one hypothesis, certain immunizations irritated tissue and predisposed the body to infection. In turn, the medical director of the Stanford Children's Convalescent Home in San Francisco, Dr. Harold K. Faber, theorized that poliovirus on the skin was driven into the body during injection and seeded into susceptible tissue.[65]

Although toxoid-based immunizations for pertussis, diphtheria, and tetanus appeared to play a leading role in polio provocation, other injections were implicated. Alarming findings were discovered by Dr. Robert F. Korns, the chief of the division of communicable diseases at the New York State Department of Health. Korns was an epidemiologist with an MD and PhD in public health from the School of Hygiene and Public Health at Johns Hopkins University.[66] His investigations revealed that injections of sedatives, penicillin, hormones, vitamins, and Novocain could also provoke polio.[67] Like Anderson, Korns based his evidence on case histories gathered from patient and family interviews.[68] Unlike earlier surveys, Korns attempted to increase the rigor of his study by including a control group comprising neighbors and family members. He concluded that common injections doubled the risk of polio paralysis for up to two months after inoculation.[69] Polio provocation appeared to have a direct relationship to a range of medical practices.

Concern about polio provocation led to significant shifts in health policy.[70] The Academy of Pediatrics Committee on Immunization and Therapeutic Procedures advised against pediatric immunizations during polio epidemics.[71] The U.S. surgeon general offered similar words of caution, acknowledging that studies showed "some risk" with administering injections during the "polio season."[72] Dr. Herman E. Hilleboe, New York State health commissioner, was especially concerned and directed county and city public health officers to discontinue "all elective immunization procedures on persons over six months" old during the summer months.[73] By 1951, most health departments and family doctors had changed their pediatric immunization practices to reduce the risk of polio provocation.

For Hammon and his supporters, the risk of polio provocation complicated plans for the GG field trial. Hammon's intention to inject thousands of children with test serums during a polio epidemic appeared to disregard Anderson and Korns's evidence and the immunization policies of many health departments. Although no studies specifically linked gamma globulin or gelatin-based placebo injections with polio provocation, the fact that similar substances appeared to facilitate infection hinted at a possible correlation.

The risk of polio provocation led the Committee on Immunization to urge caution. "I can testify that it gave us pause," Rivers later reflected.[74] Most committee members were worried by the evidence and convinced that the polio provocation posed a danger. "Some of us heard the papers of both Anderson and Korns," one scientist explained to Hammon. "I think most of us . . . were pretty well convinced that Korns' data was as good as Anderson's data." Based on this sentiment, Yale researcher Dr. John R. Paul predicted that one "might almost expect an increase in the amount of paralysis" in the injected control group. As a result, Sabin warned that Hammon "would be in a position of having produced paralysis in a certain number of children who might otherwise not have been paralyzed."[75] Although Korns's findings fueled most committee members' anxiety surrounding polio provocation, Hammon was less persuaded. "I am not too much impressed yet with Dr. Korns' data," he retorted. He reasoned that since Korns was the only investigator to have implicated a wide range of injections, his results required further scrutiny.[76]

Committee members struggled to help Hammon come up with ways to reduce the risk of provoking polio among trial participants. Some members advised replacing injected controls with oral doses of GG.[77] "There was one hell of a debate," River remembered. "It was plain to me that once you distinguished the gamma globulin from the placebo you were opening a Pandora's Box. I could just see thousands of mommas and papas of those children who got the placebo descending on the Foundation asking why their children didn't get the gamma globulin."[78] Other committee members imagined abandoning a controlled study entirely in the name of safety.[79] Although alternatives were discussed, no compromise was reached.[80] Hammon was shaken by the lack of confidence in his plan, but committed to moving forward with the experiment.[81]

Some scientists raised concerns, beyond the risk of provoking polio, about the danger of serum hepatitis contamination.[82] Hepatitis is an acute inflammatory liver disease caused by the infection of one of five hepatitis viruses (now designated A through E).[83] Symptoms range from mild to severe and can include vomiting, fever, joint pain, and yellowing of the skin and the whites of the eyes (known as jaundice).[84] Hepatitis can become chronic, leading to further complications, such as liver cancer and cirrhosis. Occasionally it can develop into fulminant hepatitis, characterized by liver failure.[85] In the 1950s, scientists only knew of two forms of hepatitis (A and B).[86] The disease could be caused by either oral-fecal exposure (infectious hepatitis) or the injection of a contaminated blood product (homologous serum hepatitis).[87] Unlike its infectious variant, serum hepatitis was less communicable and had a prolonged incubation time (60 to 150 days). It also had a higher mortality rate and was notoriously difficult to diagnose.[88] Once the illness was recognized, clinical treatments were limited to bed rest, special diets, and "physical reconditioning."[89] Even though most patients survived the acute phase of serum hepatitis, convalescence could last several months and leave lasting health consequences.

Hepatitis-tainted blood products were observed by health professionals in the early 1940s, coinciding with the application of blood plasmas and immune serums during World War II.[90] One of the most serious episodes occurred in 1942, when fifty thousand American soldiers were accidentally infected with hepatitis after receiving a contaminated yellow fever vaccine developed by the Rockefeller Institute.[91] Over eighty of those infected succumbed to the disease.[92] Despite attempts by the Rockefeller Institute and the U.S. military to allay fears, the yellow fever vaccine incident demonstrated that pooled human blood products could harbor the hepatitis virus.[93]

During the summer of 1951, public health investigators in England claimed to have evidence of hepatitis contamination in a batch of GG. Dr. W. Charles Cockburn, an epidemiologist at the Central Public Health Laboratory in London, published a report with colleagues in the *British Medical Journal* titled "Homologous Serum Hepatitis and Measles Prophylaxis." The article recounted an experiment conducted between 1947 and 1949 "into the value of gamma globulin" for "the prevention and attenuation of measles." Although the study was concerned with measles control, viral contamination of the serums afforded Cockburn and his associates "an opportunity" to investigate the role of GG in serum hepatitis transmission.[94] The researchers found one patient suffering from a mild case of hepatitis following injection of GG processed from contaminated blood serum. Based on this case, Cockburn's team cautioned that "gamma globulin can therefore be considered . . . less icterogenic than plasma or serum; but, in view of the episode here described it is probable that it is not wholly free from the risk of causing hepatitis."[95] Cockburn's article questioned the reputed safety of GG and rationalized review of existing production methods. Hammon's proposal to evaluate GG for polio therefore had to weigh the added risk of serum hepatitis.[96]

The Cockburn article worried NFIP assistant medical research director Dr. Henry Kumm. A graduate of Johns Hopkins University School of Medicine, Kumm had experience in research and had served on the Rockefeller International Health Division, the West African Yellow Fever Commission, and the Yaws Commission in Jamaica.[97] Kumm recalled the 1942 Rockefeller yellow fever vaccine disaster and was concerned that GG could harbor hepatitis. Sharing his anxiety with Dr. Harry Weaver, he stated, "I shall never forget what happened during the war when the R. F. [Rockefeller Foundation] was preparing yellow fever vaccine with normal human serum."[98] Similarly, Dr. Joseph E. Smadel expressed anxiety.[99] He sent a letter to Hammon, explaining that he was "rather disturbed" by Cockburn's findings. "This [article] at least strongly suggests that some slight risk of homologous serum jaundice accompanies the use of gamma globulin," Smadel stated frankly. "Is it great enough to influence your thinking on the poliomyelitis prophylaxis study?"[100] Couched in neutral language, Smadel's query appealed for caution.

The risk posed by large dose injections in the gluteus maximus was another worry shouldered by researchers judging Hammon's proposed trial. Since the 1940s, clinicians had learned that large or improperly administered injections in the gluteus maximus could cause sciatic nerve injury, peripheral neuritis, and muscle damage.[101] Symptoms could include severe pain and possibility of paralysis in the extremity.[102] Hammon was aware of the perils associated with large dose injections and recounted to colleagues, "I have had these large doses of [gamma globulin] serum given me intramuscularly on several different occasions, ranging up to as large as 75 cc in one buttock. . . . I have limped for two months thereafter and gone to a physiatrist to be treated for the residual effects."[103] Self-experimentation taught Hammon that the effects of large dose injections could be painful and enduring. Although most pediatric injections were administered with a 25 gauge needle (0.5 mm diameter) using less than 1 cc of serum, Hammon's protocol demanded something more impressive: an 18 gauge needle (1.3 mm diameter) using 4 to 11 cc of serum. The need for large needles to inject substantial amounts of viscous gamma globulin and placebo solution stood to increase the risk of muscle and nerve injury among child participants.[104] Since Hammon remained committed to the GG study, he and his allies imagined what improvements could be made to the protocol and how detractors might be persuaded to give their support.

Manufacturing Enthusiasm

During the 1950s, no federal mandate required Hammon to seek ethical or scientific clearance before undertaking the GG study on an open population.

Nevertheless, NFIP officials believed achieving the appearance of universal scientific approval through the Committee on Immunization was essential.[105] First, the endorsement of scientific peers would serve as proof of a viable protocol. Second, sanction would help stifle possible criticism arising from clinicians or public health officers. Third, consensus was needed for marketing and public relations, since NFIP officials believed that parents and doctors would be more willing to participate in a field trial if polio researchers were unified behind the plan.

NFIP officials were troubled by the proposed GG study's health risks and methodological deficiencies, but they nevertheless remained supportive of Hammon. The organization had invested too much time and money to allow intellectual sparring to derail a potentially valuable experiment. Between May and August 1951, they worked together to build alliances and scientific consent.[106] Among their first efforts to foster scientific agreement was to scrutinize the possible link between injections and polio provocation. Dr. Robert F. Korns, author of the most alarming study on polio provocation, was invited by the NFIP to present his data at a special gathering of the Committee on Immunization.[107] Hammon hoped to use this meeting as a forum to challenge Korns and question his findings. On July 6, 1951, the committee convened and listened while Korns reviewed earlier studies and the range of injections that appeared to cause polio paralysis. "As to how large a role this phenomenon plays," Korns concluded, "the best we can say is that it seems to be double the hazard of getting polio and double the hazard of getting paralytic polio."[108]

Although Korns's presentation upheld committee members' prior concerns, Hammon and fellow attendees questioned the findings. Hammon criticized Korns for failing to delineate his analysis by the localization of paralysis and for a possible weakness in the sampling method because of its reliance on patient interviews. He also inferred that Korns's concern over nonimmunizing agents was only tentative, since it could not be corroborated with the findings of Dr. Gaylord Anderson. Sabin continued the onslaught by critiquing Korns's implication that postinjection vulnerability to polio lasted beyond one month. "There is no logical explanation for it," Sabin reasoned, "and it is therefore difficult to understand." Rivers questioned Korns's assumption that injections were the sole culprit by arguing that perhaps the illness requiring treatment was more instrumental in causing paralysis than the injection.[109]

Although Korns's conclusions were shown to suffer from methodological deficiencies, the committee debate failed to dissociate polio provocation from the GG study. "Dr. Korns' data did not prove that there was an increased risk of getting paralytic poliomyelitis," Sabin summarized. "On the other hand, it did not disprove it. The risk remains undefined, in my mind, as it was during the last meeting, and I think we should be guided as much by the consideration of whether or not it is worth taking that undefined risk."[110] Despite uncertainty over health risks, Hammon achieved a victory by bringing into question Korns's evidence. At no time did Hammon or his colleagues recommend undertaking a separate study to assess polio provocation in relation to GG or nonimmunizing injections. Cheered by his rhetorical triumph, Hammon asserted that he saw no reason why he could not proceed with his plans.[111] By orchestrating a meeting where published research could be discredited, NFIP officials and Hammon raised a measure of doubt over other health risks.[112]

Hammon and NFIP officials salvaged the GG study by floating a compromise. "I would like to suggest that a pilot study be run this summer," Rivers counselled.[113] Many committee members concurred with this idea, reasoning that "perhaps the most leading purpose of a pilot experiment would be to try to ascertain what people will tolerate and what they will cooperate with."[114] For such attendees, a pilot study of no more than five thousand children would serve as a litmus test for public reception and an appraisal of adverse health reactions. "There will not be enough people injected to find out whether it is going to do much good," Rivers reasoned. "We are going to try to find out whether it would cause a lot of harm."[115] Although most committee members were unenthusiastic about enrolling thousands of children in a safety test, a pilot study seemed to be a tolerable concession.

Medical research tradition laid responsibility for weighing the dangers of the pilot study on Hammon as the chief investigator. In considering the decision to proceed, Dr. Joseph Smadel remarked, "I wouldn't do it myself. On the other hand, I would not interfere with anyone else doing it."[116] Within this framework, peer review existed as a means of deliberating, not policing, the ethics and utility of an experimental protocol. Since most scientists at the time considered themselves capable of balancing the risks and benefits of their experiments, any professional interference represented excessive scrutiny that threatened tradition. Although the 1947 Nuremberg Code and the 1949 Declaration of Geneva codified ethical guidelines for medical research, such doctrines were rarely observed by American scientists during the 1950s.[117] "Dr. Hammon is very modest in sitting here and having our will imposed upon him, as it were," asserted one attendee. "I think to attempt to impose something on him that he does not want to do is just wasting our time here."[118] Without the formal precedent and power of ethics review, committee members had no mandate beyond the provision of advice. Although progressive in their use of peer review, NFIP officials' decision to assemble polio researchers to review and approve Hammon's protocol proved calculated.

After reflecting on the prevailing committee sentiment, Basil O'Connor and Dr. Harry Weaver intervened to marshal approval.[119] As head of the NFIP, O'Connor had the power, experience, and personality to exert his will on committee members.[120] Since he perceived the lack of support for the pilot study as akin to a failed compromise, he demanded a united decision. "I don't look forward to our putting in the time, effort, and energy that this is going to require," O'Connor exclaimed, "unless whatever is done is done with some enthusiasm." He continued: "We just don't want to be in the position of phrasing it this way, 'Well, what about that study you are making?' 'Well, I think it is all right. Maybe it isn't. We are just doing it for the fun of it.' We can't be in that position. . . . It has got to be supported with something resembling enthusiasm or there is something wrong here that hasn't come up on the table."[121] O'Connor was frustrated by the scientific debates and lack of commitment. Unless members could explain why the pilot study should not proceed, then it must be approved. According to O'Connor, the time had come to be pragmatic, unite as a group, and support Hammon's evaluation of a blood fraction that might prevent polio paralysis before a vaccine was discovered. Following his demand for "enthusiasm," Weaver called a vote.[122] "All those in favor signify by saying 'aye,'" he announced. A "chorus of ayes" was heard. "Opposed?" No delegates responded. "I take it, then," Weaver declared, "that it is a unanimous recommendation that we proceed as outlined."[123] Casting doubt on detractors, marginalizing a potential health risk, and reducing the scale of the study made Hammon's revised protocol appear reasonable to committee members. The pressure tactics and voting scheme created the scientific approval deemed essential by NFIP officials to proceed.

Conscience and Concession

When the NFIP announced that its Committee on Immunization was united behind Hammon's field trial, some medical researchers and health officials that had not been privy to the committee debates expressed dismay. As former committee member Dr. John R. Paul remembered, the issue became "hotly argued back and forth."[124] Dr. Gaylord Anderson, who had researched and presented on the threat posed by polio provocation, emerged as a vociferous critic of Hammon's study. During an August epidemiological training session at the Communicable Disease Center in Atlanta, Georgia, he confronted Dr. Joseph Stokes Jr. over Stokes's support for the field trial. Later recounting this awkward encounter to Hammon, Stokes reported that Anderson "still strongly disapproves of our study."[125] Stokes remained convinced that proof of GG efficacy was needed to empower clinicians and that only a rigorous study could provide the evidence. Although he could not counter Anderson's concerns regarding polio provocation, he sought to dissuade antagonism by drawing attention to the inevitable clinical use of the blood fraction.

Hammon was troubled by Anderson's objections, but considered them the result of personal obstinacy. "I am sorry that Gaylord Anderson still disapproves of our study," he replied to Stokes, "but knowing him as I do I imagine he will continue to stubbornly remain of the same opinion despite all arguments." Although marginalizing Anderson's dissent, Hammon privately admitted to Stokes that such fears might turn out to be well-founded. "I hope he is not right," Hammon conceded.[126] Learning about whether GG protected against paralytic polio was for Hammon and Stokes more important that the risk of adverse reactions.

Like Stokes, NFIP officials helped Hammon check external criticism. New York State Health Commissioner Dr. Herman E. Hilleboe, who championed injection bans to reduce the risk of polio provocation, was dismayed by the planned study. Hilleboe sent a protest letter to NFIP medical director Dr. Hart Van Riper, who passed it to research director Dr. Harry Weaver for a response. Weaver understood that it was important to limit overt criticism, as it could endanger marketing the experiment. In diplomatic style, he thanked Hilleboe for his letter and attempted to assuage his concern by explaining that members of Committee on Immunization had already given the issue of polio provocation "very serious consideration." Moreover, he explained that Hilleboe's staff member, Dr. Robert Korns, delivered a report to the committee and that he might elaborate "on a number of details that I cannot go into in a letter." Weaver's reference to Korns's findings served to challenge Hilleboe and provide assurance of a careful review. Although acknowledging that pediatric injections were best avoided whenever possible, Weaver concluded that they were absolutely necessary in this case to assess "the usefulness of gamma globulin."[127]

Despite the enthusiastic backing of Stokes, Hammon appreciated the concerns raised by his scientific peers on the committee and took steps to reduce injection pain and adverse health reactions, as well as to track the extent of harm. First, he devised a standardized dosage schedule: 4 cc to children up to thirty-four pounds, 7 cc to those between thirty-five and sixty-one pounds, and 11 cc to those sixty-two pounds and over.[128] Although these dosages were large compared with the average 1 cc pediatric injection, they were the smallest

Figure 6 Dr. William McD. Hammon conducting an experiment, 1951. Courtesy of the March of Dimes Foundation.

volume of serum he could reasonably promote to parents and children, while maintaining the potential to convey protection against paralytic polio.

Second, Hammon endeavored to reduce crowd contagion by designing injection clinics that would keep "family groups isolated" and ensure the "rapid movement of children."[129] Large open floor plans combined with efficient processing of children would reduce interpersonal contact and viral spread. Although Hammon knew that it was impossible to control interactions at injection clinics, his protocol tried to mitigate the dangers.

Third, Hammon attempted to reduce and track polio provocation. Since children under two were seen to be the most susceptible to this risk, Hammon excluded them from enrollment. He also set out a rigorous regimen for clinical hygiene.[130] All packaging materials would be sterilized and all needles and syringes "autoclaved for four hours."[131] A paper sheet would be laid on the injection table and changed for each child.[132] Skin around the injection site would be disinfected with tincture of iodine and wiped clean with alcohol, which he reasoned would reduce tissue irritation and viral contamination.[133] Although he did not know whether these procedures would reduce health risks, they represented an expedient compromise. To measure polio provocation, he mandated that all injections be made in "the right buttock so that it could be determined later whether there was any association between the site of inoculation and the distribution of paralysis in children."[134]

Finally, Hammon devised a method to assess serum hepatitis contamination.[135] He created a survey comprising a detachable six-inch-by-four-inch prepaid postcard that would be sent to participating families twenty weeks after the pilot study closed. The postcard requested "data on the occurrence of several common acute communicable diseases and 'yellow jaundice'" among children. In cases where a returned survey hinted at a positive case of hepatitis, Hammon pledged to send medical assistance. "All jaundiced cases will be followed up by a visit," he assured NFIP officials "from a physician member of our team to get further details." Although the postcard and its associated follow-up program demonstrated some diligence in tracking the potential health consequences of the field trial, it also revealed a dubious resolution to the lingering doubt over serum hepatitis.[136]

The postcard combined public relations acumen with research imperatives. It featured a brief introduction and questionnaire that could be folded over and mailed after completion. Hammon phrased the preamble to present the card as a routine communiqué rather than a method to track serum hepatitis. "We are attempting to determine whether gamma globulin may have given protection against several infectious diseases," the opening section stated.[137] Even though the pilot study was never intended to measure the wider prophylactic value of GG, Hammon invoked such language to cloak the true intent behind the form.[138] The postcards asked parents to confirm their child's name, birth date, and inoculation date, as well as indicate whether they had "suffered from the following specific [list of] diseases" itemized on the form.[139] Common illnesses were listed, including influenza, chicken pox, measles, mumps, and

Dear Parents:

Our records indicate that your child, who is named on this postcard, was inoculated in the Poliomyelitis Clinics in Sept., 1951. We are attempting to determine whether gamma globulin may have given protection against several infectious diseases and also whether we have complete records in regard to all cases of poliomyelitis which may have developed. Would you please mark an 'x' in the appropriate block opposite the name of any disease listed on the attached return postcard which your child may have developed between the date of inoculation and the time you receive this card. Kindly list also the exact or appropriate date of onset of such diseases and the name of the doctor in attendance, if any.

We would appreciate greatly if you would give this your immediate attention and return the attached portion of the postcard by mail. No additional postage is required.

Very sincerely yours,
W. McD HAMMON, M.D.
Director
Poliomyelitis Project, Utah County

______________________________ Date of Birth ______________
(Name of child) (mo. day year)

Since September ____, 1951, when this child was inoculated in the Poliomyelitis Clinic, __ he has suffered from the following specific diseases:

	Mark X in Block	Date of Onset
Chicken Pox	☐	
Measles (regular)	☐	
German measles	☐	
Mumps	☐	
Poliomyelitis	☐	
Scarlet Fever	☐	
Yellow Jaundice	☐	
Whooping Cough	☐	

Name of physician, if any, in attendance of above illness:

____

____

Figure 7 Hammon's hepatitis postcard questionnaire, 1951. Courtesy of the March of Dimes Foundation.

whooping cough. Serum hepatitis was not specifically indicated among the options; instead "yellow jaundice," one of the clinical manifestations of hepatitis, was listed. The placement and terminology chosen for serum hepatitis showed that Hammon did not want to rouse any concern among participating families.

Although convenient and economical, the hepatitis postcard had distinct limitations. The questionnaire assumed that all diseases identified by the parent were attended to and diagnosed by a physician. Though medical advice would likely be obtained in the most serious cases, there was no guarantee that every illness had an accurate diagnosis; serum hepatitis was notoriously difficult to identify and could manifest itself in symptoms different from "yellow jaundice." Mild cases of serum hepatitis could also go unobserved and unrecorded. Moreover, Hammon assumed that the postcard would be completed accurately and returned promptly. Despite its limitations, the postcard survey pleased NFIP officials. Kumm was especially relieved that a means to track the risk of serum hepatitis had not been disregarded by Hammon.[140] "I am very glad," he wrote, "that you have a question about homologous serum jaundice." He concurred that the postcards be "sent out to the parents about four months after the inoculations were given," since this duration exceeded the average incubation period of serum hepatitis.[141] Although the postcard could provide some indication of hepatitis among the cohort, it was far from comprehensive.

Health risks pervaded Hammon's proposed pilot study, but his belief in the need to assess the efficacy of GG for polio allowed him to remain undeterred by criticism. Through a standardized serum dosage schedule, revised clinic design, cautious injection procedure, and postcard survey, he attempted to reduce the harm his study might cause, while also tracking its consequences. Just how much information would be shared with families about the risks of the study was subject to further consideration. For Hammon and NFIP officials, the biggest test was yet to come: how to market the experiment to the public.

Chapter 3

Marketing and Mobilization

"I am sure you are equally appreciative of the potential dangers if this whole field trial is not handled in the best possible way," National Foundation for Infantile Paralysis (NFIP) medical director Dr. Hart Van Riper reminded Basil O'Connor in June 1951.[1] While the Committee on Immunization had given its support for the gamma globulin (GG) pilot study, the NFIP was clearly aware of its role in seeing the study come to fruition, and Van Riper was anxious about the next step: how to sell it to the American public.[2] Such an endeavor would not be simple, since a large placebo-controlled study had not been undertaken on an open population. The scale would be tremendous, featuring the rapid enrollment of thousands of healthy children in temporary injection clinics at the height of a polio epidemic. Coming up with a strategy for compelling parents to volunteer their charges for a study that did not promise protection would test the best marketing minds. For these reasons, Van Riper recommended a "carefully projected public education campaign" to make the experiment palatable.[3] Faced with an uncertain public reception, NFIP officials devised an ambitious promotional strategy. However, differing expectations and approaches led to challenges in launching the study. What preparations were made to ensure the successful launch of the pilot study? How would marketing shape the study's nature and delivery? This chapter explores how preparations for the experiment were complicated by differing agendas and what steps were taken to ensure its launch.

Preparing for a Pilot Study

In the spring of 1951, Hammon began enlisting a field team comprising epidemiologists, medical researchers, clinicians, and orthopedic nurses. The core team was made up of close friends and associates. Hammon appointed tireless GG promoter and ally Dr. Joseph Stokes Jr. as senior trial consultant. He also recruited Dr. Lewis L. Coriell, who was medical director of the Camden Municipal Hospital in New Jersey, to serve as deputy trial administrator. Coriell came from a prominent medical background; he held an MA and PhD in microbiology, as well as an MD from the University of Kansas.[4] In addition to Coriell, Hammon's colleague, Dr. Francis S. Cheever, was chosen to assist with the trial because of his background in microbiology and infectious disease control.[5] As trial administrators, Hammon, Stokes, Coriell, and Cheever shared considerable experience with polio, knowledge of medical research methods, and belief in the need for a GG study.

Figure 8 Planning the gamma globulin field trial, 1951. Pictured (left to right) are Dr. Lewis L. Coriell, Dr. William McD. Hammon, Dr. Joseph Stokes Jr., and Dr. Henry Kumm. Courtesy of the March of Dimes Foundation.

With a team assembled, Hammon finalized his study's protocol. Over five thousand children would be enrolled in a study that adhered to controlled trial methodology. The placebo and GG serum doses would look identical at room temperature; however, each dose would be assigned a unique tracking number, which would be recorded in a schedule before being randomized. Half of the pediatric cohort, ages two to eight years, would receive a 4, 7, or 11 cc injection of GG based on body weight, while the other half received an equal dose of gelatin solution. The similar appearance and random allocation of the serums at the time of injection would prevent anyone from knowing which child received which dose. Once the study was finished, the dose tracking numbers would be used to compare the two groups to determine whether GG provided any protection against paralytic polio.[6]

As a large controlled experiment on an open population of healthy children had never been attempted before, trial administrators preferred to undertake the study in a setting with a specific racial, cultural, and geographical composition. Although a large city was an obvious choice for an experiment, with a high population density and access to trained medical personnel, transportation corridors, and clinic facilities, trial administrators believed that an urban context was inherently unsuitable; instead, their insecurity about public reception led them to seek the assistance of a small town.[7]

A small town was suitable for an experiment for many reasons. Trial administrators believed that such a community would be more socially cohesive than an urban metropolis. Evidence of cultural pluralism, as well as simmering racial, ethnic, and class divisions situated many urban centers as disunited communities.[8] Since thousands of children needed to be enrolled in a short period of time, community division stood as a threat to public relations. "If we go into a major city of five hundred thousand or a million people and try to sell the [pilot study]," Hammon warned, "that may be something altogether different" than "getting the cooperation of a small town."[9] Informed by a culture of interdependence, shared affiliations, and solidarity, small towns appeared ideal for generating intense approval.[10]

Hammon also believed small-town doctors would acquiesce to medical researchers at a time of crisis. Unlike most urban doctors working in large hospitals, small-town doctors lived in the communities where they practiced and often nurtured friendships with their patients.[11] The emergence of polio epidemics in sparsely populated regions brought considerable strain to local doctors, many of whom felt powerless to protect the children they had helped deliver into the world.[12] Evidence also suggests that Hammon believed small-town doctors were less informed of current medical issues than their urban counterparts.[13] "You usually have a somewhat different group of physicians in the relatively rural communities and the smaller towns," Hammon reflected, unlike urban doctors, who would "have to be sold on this experiment before they [would] endorse it."[14] He further reasoned that small-town doctors would not deviate from the protocol; in particular, he was deeply troubled that the study might be ruined if doctors caved to parental pressure and administered clandestinely acquired GG to local children. Since Hammon knew that most small-town doctors rarely stocked GG or knew about its potential application for polio, their capacity to undermine the protocol appeared limited.[15] The ostensibly rustic culture of small-town America and its emphasis on family medicine was presumed to offer a special environment that fostered compliant physicians that could bridge the scientific and domestic worlds.

A rural small town also promised a measure of operational privacy. Since the GG pilot study was an evaluation of safety, logistics, and public relations, trial administrators reasoned it was advantageous to locate the project away from medical detractors or prying journalists. This was important for two reasons: first, Hammon wanted to keep awareness of GG quiet until efficacy against polio was confirmed; second, he did not know how the study would turn out. If parents failed to volunteer their children or if adverse health reactions were observed, it appeared wise to restrict knowledge of such developments. "There are small communities where things like that [a GG experiment] could be done without any fuss or fear at all," Michigan polio researcher Dr. Thomas Francis Jr. informed Hammon. "Now there was a town that had a population of approximately 2,000 and we were able to get [fecal samples] from all children under sixteen years of age, and I don't think anybody outside the town knew that we were doing anything."[16] By favoring a small-town context, trial administrators perhaps intended to re-create the "wall of silence" prevalent in institutionalized medical experimentation.[17] Such rationale mirrored atomic weapons testing at the Nevada Test Site in southeastern Nye County, Nevada; CIA experiments with the powerful hallucinogen LSD in the village of Pont-Saint-Esprit, France; and the study of untreated syphilis among African American men in Tuskegee, Alabama.[18] A small-town proving ground assured a measure of sovereignty that Hammon believed necessary for a study with an unknown outcome.

A small town was also a practical compromise between the privacy assured by a sparsely populated region and the resources of urban settlement. At a minimum, trial administrators required large public buildings to use as improvised injection clinics and access to developed transportation corridors.[19] Since all emergent paralytic polio cases within the test site would be assessed throughout the epidemic by a follow-up team, it was crucial that hospitals with trained personnel and suitable equipment were available. The alternative of improvising a makeshift tent clinic, as was established during the 1944 polio epidemic in Hickory, North Carolina, was deemed too expensive to staff and equip.[20] Small towns assured the minimum operational standard to manage a large medical experiment.

A small population base would also help restrain public enthusiasm. Since the pilot study would be undertaken with a limited quantity of GG and placebo solution, it was important that the number of potential child volunteers did not overwhelmingly exceed the number of available doses. Should the experiment prove popular among parents, Hammon imagined that serum supplies would be quickly exhausted, forcing possibly hundreds of families to be turned away from the clinics.[21] He feared that disappointment would pose a public relations problem and encourage some parents to seek a private supply of GG, thereby damaging the statistical underpinnings of the experiment. Selecting a community with the correct pediatric population assured some measure of operational control.

Perhaps most importantly, trial administrators required the test site to be in the throes of a high-incidence early-stage polio outbreak with indications of lasting intensity. Hammon theorized that by commencing the experiment at the onset of an epidemic, rather than at its peak, the protective antibodies in GG would have more time for absorption and opportunity to fend off paralysis. The more children injected prior to the highest incidence of infection, the more likely that polio protection (if any) would be realized. Moreover, since the case rate of paralytic polio was quite low (approximately 25 per 100,000), only a virulent and lasting outbreak could assure a suitable environment for statistical comparison.[22] Epidemic phase, degree, and longevity were elements that Hammon weighted when identifying a test site.

Creating a Medical Marketing Program

Complementing Hammon's scientific preparations, NFIP officials began planning a program to promote the experiment. Members of the Committee on Immunization were tasked with helping Hammon and, after considering the numerous challenges, endorsed marketing as a means to build public support for the study. "Any program necessitating the cooperation of twenty thousand to forty thousand children in receiving injections," observed Indiana health official and committee member Dr. Leroy Burney, "will be difficult but not impossible."[23] Since a large public medical experiment was unprecedented, some committee members foresaw the study as a useful indicator of what people would tolerate. "It seems to me," mused Children's Hospital of Boston polio researcher Dr. John Enders, that "this experiment . . . is to test out the public relations and whether it can be done from that point of view."[24] NFIP scientific adviser Dr. Thomas M. Rivers concurred with Enders, asserting, "I should think it might be run largely to see . . . what your public relations problems are."[25] Uncertainties about public reception of the study caused committee members to urge Hammon and NFIP officials to develop a sophisticated publicity program.

Publicity would be important to the trial, but it would be complicated to implement since there was lack of advance notice to prepare the target audience.[26] Unlike the March of Dimes fund-raising campaign, which benefited from weeks of promotion before donations were solicited in January, marketing for the field trial would be restricted to hours or days.[27] As the enrollment of children needed to begin soon after the test site was selected, only a well-organized and aggressive publicity program could rouse a community into action.

Anticipating and assuaging parental concerns about health risks would also be an important facet of publicity. The danger of crowd contagion was part of public consciousness, due to newspaper features and NFIP literature.[28] As a result, the study needed to be presented as safe and necessary in order to counter advice that proclaimed crowd avoidance was the best means to guard against polio.[29] Moreover, physicians needed to be assured that neither GG nor placebo injections would cause polio, since some were cognizant of the polio provocation evidence and had stopped pediatric injections during outbreaks.[30]

The public also needed accessible information about the methodology for controlled field trials.[31] Even though researchers had long employed observed controls for comparative purposes, few parents or doctors were familiar with a double-blind placebo-controlled field trial. According to medical historian Harry Marks, "As late as 1950, most physicians still thought of statistics as a public health domain concerned with records of death and sickness."[32] Since not all Americans understood evolving field trial methodology and the value of medical statistics, public education was needed to assure support for the experiment and adherence to its protocol.[33]

As parents had the freedom to choose whether to volunteer their children, on what day to enroll them, and whether to abandon the study, marketing would play an important role in achieving a sufficient cohort.[34] Since it was difficult to know how long a polio epidemic would last or when it would peak, the faster the maximum cohort was enrolled, the greater the opportunity for a valid assessment of GG. Hammon needed families to respond swiftly and in a steady, orderly stream. "We are going to inject volunteers," Hammon related, "frequently several children in the same family, and we must keep them coming in large numbers over a period of days until thousands have been injected."[35]

Criticism or apprehension could lead to underutilized clinics, while overoptimism and desperation might inspire long lines. Appropriate publicity promised to manage the public response.

The NFIP assigned the role of managing publicity for the GG study to public relations director Dorothy Ducas. Born to a prosperous New York family and a graduate of the Columbia School of Journalism, Ducas was an experienced, well-connected, and savvy publicist. Her professional career began in 1927 when she was hired by the *New York Evening Post* as a reporter. "I did features and news. In fact, being a reporter was the best part of my life," Ducas remembered. In 1928 when Franklin D. Roosevelt was running for the governorship of New York State, Ducas met and interviewed his wife, Eleanor, who "was then working for the Democratic State Committee." The interview proved momentous for Ducas, as it was the beginning of a lifelong friendship.[36] With the onset of the Great Depression, Ducas left the *Post* to pursue a position as editor of the women's magazine *McCall's*. She wrote on such subjects as Prohibition repeal and the lives of celebrities.[37] Following cutbacks, Ducas left *McCall's* to join the major Hearst news agency International News Service (INS). During her tenure at INS, Ducas met Tom Wrigley, the publicity director for President Roosevelt's initial polio charity, the Committee for the Celebration of the President's Birthday (CCPB). Ducas recalled her first meeting with Wrigley as gratifying: "He admired my work and I admired his." Wrigley recognized that the CCPB needed someone to mobilize women's groups and invited Ducas to serve as the director of women's publicity. Ducas did not accept his offer immediately, but sought the opinion of Eleanor Roosevelt, who "said she thought it was a wonderful idea because the [organization] needed somebody like [Ducas]." It was at this juncture that Ducas migrated from journalism to embrace a new career as a charity publicist.[38]

When Ducas began working with Wrigley in October 1937, she helped reorient polio fund-raising away from the politically charged CCPB to the ostensibly apolitical March of Dimes. As public relations educator Scott M. Cutlip quipped, "Wrigley and Miss Ducas quickly shifted the main appeal from one of paying tribute to the nation's President to one of unashamed exploitation of the pathetic appeals of crippled children."[39] Ducas excelled at a range of duties during the formative years of the NFIP. She remembered that her job was "to get hold of people who were influential and get stories from them, also to interview patients and their families and to contact all the magazines—particularly the magazines because I had been a magazine editor and writer—and get them to give space to the March of Dimes in their January issues—our month for collecting money."[40] Ducas also organized and chaired special March of Dimes events for the purpose of soliciting funds and recruiting women volunteers. She and Eleanor Roosevelt held the Women's Week of the Fight Infantile Paralysis campaign at the White House, in which prominent socialites participated.[41] Ducas rapidly became influential at the NFIP and successfully promoted the fight against polio to the nation.

In 1942, Ducas joined the Office of War Information (OWI), an agency established to promote the war effort at home and to consolidate government publicity.[42] Under the direction of former *New York Times* journalist and CBS news anchor Elmer Davis, Ducas became director of the OWI Magazine Bureau. Through this division, Ducas and her staff carried out an ambitious public relations campaign to increase the quantity of vetted information reaching news editors and minimize public misunderstandings about government policy.[43] Ducas remained with the NFIP and by 1944 was reassigned by Basil O'Connor to the role of "special public relations consultant." Due in part to Ducas's accomplishments at the OWI and O'Connor's confidence in her abilities, she was invited in 1949 to become the NFIP's director of public relations.[44] The promotional engine of the NFIP during the 1950s was fundamentally the realm of Ducas, a person with a journalist's instinct for a story and a keen understanding of methods to sell the NFIP to the public.

O'Connor and Ducas believed that the NFIP was best suited to spearhead the promotion of the GG study. Considering the scale and time restrictions involved with the experiment, they reasoned that NFIP involvement would guarantee a favorable public reception and assure recognition for their investment. As the primary economic sponsor of the study, the NFIP had allocated $100,000 for researcher salaries, medical contractors, serum bottling, and supplies.[45] This was a substantial sum when a bread loaf cost sixteen cents and an automobile could be driven home for $1,500.[46] By taking control of publicity, NFIP officials hoped to steer promotion away from amateur mistakes.

Hammon was irked by NFIP officials' plans to manage publicity for the field trial and offended that they had not included him in preliminary marketing meetings. "I appreciate your reasons for feeling that the Foundation needs to be safeguarded in this publicity and that not getting my ideas earlier was not done because of lack of confidence in me," he wrote to NFIP research director Dr. Harry Weaver. "But, in case some apprehension was felt in respect to my ability along these lines I would like to point out that I have had considerable formal training in health education and in selling health projects to the public. . . . I feel a certain right to believe that I know something of the art."[47] Hammon was concerned that the association of NFIP personnel might undermine his experiment's aura of scientific neutrality and transform it into a publicity stunt for the March of Dimes. "I am hesitant about notifying the public relations office of The National Foundation," he explained to Weaver. "I admit I do not know your personnel, and all my dealings with their representatives took place several years ago."[48] Although Hammon valued the financial and logistical backing of the NFIP, he was unconvinced that the organization should also manage public relations.[49]

Hammon articulated his own vision for marketing the GG study. He believed that glossy prearranged publicity was unsuitable and that an impromptu program would be easier to manage. "The Foundation is going to have to allow me to ad lib as necessary," he insisted.[50] Hammon was confident that the experiment would sell itself; he expected that once the state and county medical societies approved the study, it would be a small matter to attract parents.[51] Akin to Dr. Sidney Kramer's attitude years earlier, Hammon presumed that the crisis of an epidemic would be sufficient impetus for parents to enroll children.[52]

Although NFIP officials agreed that a polio epidemic would inspire some parents to volunteer their children, they believed that Hammon had underestimated the degree of publicity required. To appease Hammon and advance their interests, NFIP officials negotiated a compromise in which their personnel would appear incognito at the test site and not draw attention to the March of Dimes.[53] Weaver assigned to the project his subordinate, assistant research director Dr. Henry W. Kumm, to liaise with Hammon to assure cooperation.[54] The compromise also included a promise to shield Hammon from direct encounters with Dorothy Ducas and her public relations staff by hiring New York journalist Robert B. Sullivan as an intermediary marketing consultant.[55] A resident of Bedford Village and a contributor to the Sunday Section of the *New York News*, Sullivan was an apt choice, as he had experience preparing publicity and was comfortable translating complex ideas into accessible language.[56] His assignment was to liaise between Hammon and the NFIP Committee on Immunization's subcommittee on publicity, give advice, and craft media releases.[57] In exchange, Hammon agreed to allow NFIP personnel to manage publicity and design marketing materials under the supervision of the subcommittee. Through this compromise, Hammon and NFIP officials agreed to collaborate on how the GG study would be sold to Americans.[58]

Bound up with public relations challenges were legal issues. State medical licensing laws mandated that only physicians licensed to practice medicine could administer injections to children.[59] As the experiment was unlikely to occur in a state where Hammon or his team was licensed, local doctors would need to carry out the injections. Furthermore, parental consent was required to authorize the enrollment of children. Finally, as participation in the experiment did not guarantee protection from polio paralysis and had potential health risks, complications needed to be anticipated and a legal waiver adopted to protect scientists and the NFIP.

NFIP officials retained William F. Martin, of the firm Martin & Clearwater, New York, for legal counsel.[60] Martin & Clearwater was a highly regarded agency with experience representing medical institutions; Martin himself served as counsel to the New York State Medical Society.[61] The engagement of Martin revealed NFIP officials' desire not only to conduct the pilot study in a lawful manner, but also to harness the law for their own ends.[62] Through a series of private meetings held at NFIP headquarters, Hammon, NFIP officials, and a team of researchers, lawyers, and writers set about crafting a suitable marketing plan.

Martin recognized that conducting the field trial would be difficult. He explained to NFIP officials that Hammon and his colleagues were also "well aware of many of the problems" involved with taking a study to an open population. "Of course the principal one would be the natural reluctance of parents to allow any injection of a child unless the parents are convinced that it will accomplish a tangible therapeutic result," Martin reasoned. He understood that Hammon's study would not guarantee polio protection; instead, the study only aimed "toward adding to the general knowledge of the causes of and remedies for infantile paralysis by checking on the children who are injected, over a period of many months." Promising parents a 50 percent chance that their child would receive a dose of a substance with an unknown efficacy appeared to Martin as an insufficient enticement. "It would therefore seem that the whole project is doomed from the start," he remarked.[63]

With the assistance of Martin, Hammon and NFIP officials devised a hierarchical marketing strategy to generate popular support for the pilot study. The idea of shaping public opinion through a calculated strategy was not novel, but was an established practice evident in most political campaigns, product advertising, and fund-raising. Prominent journalist Walter Lippmann, in his famous treatise *Public Opinion,* examined how the "manufacture of consent" could be achieved in an American context.[64] Whether it was tobacco company R. J. Reynolds implying the healthful benefits of cigarettes under the slogan "More Doctors Smoke Camels" or Ford Motor Company describing its Victoria model as "the Belle of the Boulevard," the idea of seeding ideas through slick promotional campaigns and the testimonials of prominent citizens connected to a long history. As a publicly funded charity, the NFIP was successful at popularizing the fight against polio and the need for contributions through celebrity endorsement and sophisticated marketing. By drawing on this vast experience, NFIP officials hoped to make Hammon's experiment palatable to cautious doctors and parents.[65]

They imagined segmenting American social and professional hierarchies into distinct subgroups: polio researchers, state health officials and clinicians, county health officials and clinicians, civic officials, the media, and, finally, parents.[66] Starting at the top of the hierarchy, each group would be encouraged to approve the experiment based on prior evidence of assent. So long as each group recognized its position in the hierarchy and trusted to the legitimacy of higher strata, a potent mechanism to shape opinion would be established. The process would be concealed so that only when overwhelming support encompassed all higher strata would parents be notified.

The market segmentation strategy also promised to protect the experiment from criticism and mobilize local resources. The approval of the most influential group, polio researchers on the NFIP Committee on Immunization, would help legitimize the project against criticism from wary health professionals. The sanction of the subsequent group—state and county doctors—would provide important intelligence about the proving ground. In the event that state-level agencies advised against the plan, little time would be lost and the search for another site could resume.[67] Local doctors were also important for legal and public relations reasons.[68] Only doctors licensed in the state where the study was conducted could legally administer or supervise injections. Moreover, the influence of doctors within their communities was considerable.[69] The parallel growth of scientific motherhood, where parenting was no longer seen as an innate quality but a learned skill, strengthened the prestige of doctors.[70] Many mothers turned to manuals and physicians to adapt child health to scientific models; immunization programs were seen by many as evidence of responsible motherhood. The influence of local doctors on civic and school officials would assure access to volunteers and buildings in which to house the injection clinics.[71] The hierarchical system would exploit social networks and local resources to make the experiment possible.

Hammon and NFIP officials' conception of medical ethics shaped how publicity about the GG study was fashioned and disseminated. Prior to the establishment of Institutional Review Boards in the 1970s, the practice of allowing chief investigators to devise their own experimental protocols and their own standards of practice prevailed.[72] Although most researchers shared a loose understanding of what was ethical in the pursuit of medical knowledge, the absence of institutional oversight allowed for subjective interpretations and justification for a range of experiments, including the Tuskegee syphilis study, the Holmesburg Prison study, and the Cincinnati radiation study.

Following World War II, widespread awareness of Nazi medical research atrocities renewed concern over experiments conducted on human subjects. After the famous Nazi Doctors Trial of 1947, the Nuremberg Code was published, providing guidelines on how medical experiments should be conducted. The code mandated that human subjects have the "legal capacity to give consent" and "exercise free power of choice" based on "comprehension of the elements of the subject matter involved" and "all inconveniences and hazards reasonably to be expected."[73] Although a few American medical researchers perceived the code as a turning point, most remained ignorant of its directives or believed them only applicable to researchers from former enemy nations.[74] The indifference most American researchers held toward the code provided the context from which Hammon and NFIP officials pursued consent for the GG study.

To systematize enrollment and protect the experiment from litigation, Hammon, Martin, and O'Connor created a parental consent form.[75] As lawyers, O'Connor and Martin knew that obtaining the signed permission of a parent was important, since children were minors and could not enter into a binding contract by themselves. Their decision to obtain consent by proxy was part of a long-standing tradition; in 1900, for example, Chicago physician W. K. Jaques administered therapeutic injections to two boys suffering from scarlet fever after obtaining the "consent of the father."[76] The GG study form of consent was a type of legal waiver, asking parents to authorize that their children receive a random allocation of injected serum to assess a possible means of preventing paralytic polio. Hammon was listed foremost in the opening statement, with "financial support" attributed to the NFIP, followed by a cascade of six supporting groups, including the Committee on Immunization as well as the local medical society and health department. This impressive array of official organizations extended credibility to the study and helped allay parental concerns.[77] Since Hammon and NFIP officials were concerned with public perceptions, their rationale for consent was linked to marketing and legal agendas. Potential health risks were not acknowledged or enumerated. Moreover, the decision to distribute the consent form to frightened parents under tight time constraints rendered immaterial the notion of informed consent.

To control the flow of information about the GG study, NFIP officials created a media relations program. Whereas medical experiments conducted on institutionalized populations were insulated from media attention, the GG pilot study would be open to scrutiny. Journalists, photographers, and radio personalities were either potential allies who might endorse the experiment or enemies who might criticize the venture. Such intrinsic fascination would

FORM OF CONSENT

I have been informed that Dr. William McD. Hammon, Professor of Epidemiology of the Graduate School of Public Health of the University of Pittsburgh, with the financial support of the National Foundation for Infantile Paralysis, Inc., and with the cooperation of the .... County, and .... State Health Departments, has inaugurated in .... County, ... a project which aims at the injection of a large number of children for test purposes in connection with the study of Infantile Paralysis and its causes and remedies. For the purpose of that test, it is proposed that a solution of gamma globulin or a solution of specially prepared gelatin be injected into a buttock of each of the children. The method of identifying the substance injected into each child is such that the person administering the injection does not know which one is being used at the time. To assist this project, I hereby consent that either one of the preparations above mentioned may be injected into the buttock of my child

....

(Name of Child)

Signed: ....

(Father or Mother)

....

(Relationship to Child)

....

(City or Town)

Witness: ....

Date: ....

Figure 9 Form of consent for the gamma globulin field trial, 1951. Courtesy of the March of Dimes Foundation.

inevitably draw commentary; as Dorothy Ducas warned, "It is the sort of story newspapers 'go for.'"[78] At the local level, media coverage had two agendas: the study had to be attractive enough to entice parents, but tempered enough to avoid implying that GG was effective. At the national level, the coverage had to address widespread interest, yet discourage parents from seeking private GG injections from family physicians.

The NFIP hoped to shape reporters' interpretations of the experiment using prepared publicity. With the assistance of Robert Sullivan, NFIP officials wrote special background stories and "approved" news releases for distribution when the study commenced. Such statements were vetted by the Committee on

Immunization's subcommittee on publicity to certify that they conformed to the scientists' agenda.[79] Prepared scripts furnished biographical information about the trial administrators, described the controlled field trial methodology, and asserted the safety of the test serums.[80] "Half of the children will be inoculated with a human blood fraction," one prepared news story explained, "which has shown encouraging results in controlled experiments on poliomyelitis with animals, and in uncontrolled tests with human beings."[81] A complementary radio broadcast script clarified that "the test does not promise protection from polio."[82] By anticipating media demands, NFIP officials hoped to discourage sensationalist coverage and shape how the study was presented.[83]

The NFIP hoped to mediate the interaction between Hammon's team and journalists.[84] To reduce misstatements, Dorothy Ducas advised the team to address journalists from a prepared script and "be willing to repeat [the content], even if word for word."[85] She cautioned that failure to follow the script might incite reporters to make "up their own stories, with embarrassing results." To deflect difficult questions, she recommended that scientists use "no comment," since journalists "understand the reticence of scientists."[86] Ducas believed that trial administrators could exploit popular assumptions about medical scientists as inherently cautious, precise, and aloof professionals wary of journalistic hyperbole. Although Dr. Harry Weaver concurred, he reasoned it was better publicity for scientists to evade unwanted questions by stating that they had "absolutely nothing further to report than that already confirmed in the [news] releases."[87] By championing adherence to scripted materials, NFIP officials fostered a cautious approach toward media interactions.

As well as press statements, the NFIP was keen to control photographic content. Whereas published pictures of children with smiling families would lend credence to the experiment, pictures of long needles might convey negative sentiments. At a time of widespread photo journalism, Ducas reasoned that trial administrators needed to devise a method for "handling" local and syndicate news photographers "during the injection period." Although her recommendations did not explain what "handling" entailed, evidence suggests that it pointed to restricting access. Since the NFIP did not want to be seen as obstructionist, Ducas and Weaver recommended that health officers might be enrolled to "handle this end."[88] For the NFIP, the press was a dubious ally that required close supervision.

In addition to preparing publicity, NFIP officials worked to fortify their public relations system. Depending on the goodwill of journalists for a successful March of Dimes campaign, NFIP officials believed it was prudent that their staff not appear evasive when responding to media inquiries. Ducas reminded

O'Connor that reporters will "not accept kindly" the "no comment" response from NFIP staff, whom they considered "co-workers."[89] Instead, she reasoned that the organization needed to prepare for its encounter with the press and conceive of responses to "special and unforeseeable questions."[90] A public relations training program was instituted at headquarters to instill vigilance and standardize responses. Weaver prepared a series of statements that personnel were expected to memorize. He identified six troublesome questions with corresponding "responses." The prescribed approach was to remain noncommittal, yet evoke a sentiment of openness. "You have been told every single thing that we know," his suggested reply began. "When there is anything new released, that will be released to you as requested." For "persistent news or radio reporters," who would not accept stock statements, Weaver set out a rhetorical strategy that contrasted the supposed muckraking tendencies of journalists with the sincerity of the NFIP. "We have been absolutely and completely above board with you," the scripted response affirmed. "These [press] statements have been gone over with the view to giving you only what is absolutely true." Although there was a considerable amount of information concerning the pilot study that would not be disclosed, evoking the notion of transparency eroded the grounds for persistent inquiry.[91] By devising mechanisms to manage inquiries, NFIP officials hoped to maintain good press relations and reduce sensationalism. With a comprehensive publicity program in development, attention turned to the logistical side of the experiment.

Launching the Experiment

The production phase of the pilot study started without complications; the NFIP and the American National Red Cross (ARC) selected reputable firms to prepare the serums. Knox Gelatin Company of Johnstown, New York, was contracted to supply gelatin for use in the placebo. The Knox family had developed granulated gelatin for commercial and private applications since the 1890s, and their company was a prominent supplier.[92] E. R. Squibb & Sons Ltd. of Brooklyn, New York, was chosen to lead production and prepare the Knox sterile gelatin and ARC gamma globulin serums. Founded by Dr. Edward R. Squibb in 1858, the company marketed a range of popular pharmaceuticals and was experienced working with GG.[93] The Squibb blood program, with its earlier investment in the Cohn fractionation method, was considered among the most advanced.[94] Experienced suppliers promised Hammon and NFIP officials a quality product delivered in a timely manner.

American food and drug regulations assured a measure of federal involvement in the experiment. Although the 1902 Biologics Control Act, the 1906

Pure Food and Drug Act, and the 1938 Food, Drug, and Cosmetic Act defined an active role for the federal government in pharmaceutical regulation, the Food and Drug Administration and National Institutes of Health had no jurisdiction over the conduct of medical experiments that used serums administered without fees.[95] However, the federal government had the power to regulate the production and sale of GG. After the 1942 Rockefeller yellow fever vaccine disaster, the Division of Biologics Control, which served as the laboratory arm of the National Institutes of Health, was tasked with overseeing GG production standards.[96]

Among the initial challenges for Squibb personnel was to develop a safe placebo solution derived from Knox gelatin. Physical chemist and production chief Dr. J. Newton Ashworth formulated the serum after tests at the New Brunswick laboratory. As a former student of blood fractionation pioneer Dr. Edwin Cohn, Ashworth settled on a 1.5 percent gelatin in aqua solution with 0.9 percent sodium chloride to act as a preservative.[97] Although injected gelatin was later observed to cause allergic reactions, such occurrences were unknown at the time.[98] If properly prepared, gelatin injections were presumed to be innocuous.[99] Ashworth's placebo solution looked nearly identical to GG at room temperature, but he admitted that refrigeration led to observable differences. Gelatin, unlike GG, was much more reactive to temperature and reached a higher viscosity when cooled. However, since gelatin was presumed among the safest naturally viscous compounds that could be injected into humans, there appeared little choice but to accept this shortcoming. To mitigate the problem, Ashworth advised Hammon "not to keep the GG and gelatin solutions under refrigeration" during the study.[100] The deliberate absence of refrigeration was not considered a danger, as the biological substances would be utilized almost immediately following processing. In spite of its shortcomings, the placebo met the basic requirements of trial administrators.

Despite encouraging early developments, the failure of Hammon, the ARC, and the NFIP to assign a project manager to coordinate the logistical facets of the experiment led to confusion and production delays. Just as "big science" programs such as the Los Alamos Manhattan Project necessitated a centralized plan to assure that theory and practice converged at the right time and in the correct order, so too did a large medical experiment require a planning authority.[101] The problems began in April 1951, when Hammon and NFIP officials announced plans to undertake a GG experiment by the early summer. In preparation, Dr. Sam T. Gibson, associate medical director of the ARC, requested that a supply of dried GG be allocated to the project and reconstituted at Squibb. However, when members of the NFIP Committee on Immunization voted in

May to halt the project until safety concerns were resolved, the liquefied GG remained unprocessed in large bulk glass containers. Once the committee eventually agreed to approve a pilot study in July, Squibb personnel were reportedly not informed of this decision; instead, Ashworth received a series of baffling letters throughout July and "a number of conflicting instructions," which he considered evidence of sustained scientific discord and indecision. Bottling of the GG was not resumed. Meanwhile, Hammon and NFIP officials wrongly surmised that the serums were bottled and would be ready for release by early August. It was not until late July when Hammon discovered and clarified the misunderstanding that production was reinitiated.[102] Failure in key aspects of production management had cost trial administrators valuable time.

Since the GG assigned to the pilot study was prematurely reconstituted, concerns about its safety were raised. Lengthy bulk storage of the blood fraction was not normally allowed, since it increased the chance of protein denaturation and microbial growth. By July 1951, Ashworth was concerned that the GG had "developed pyrogenic [fever-inducing] properties" after months in storage.[103] Since discarding the expensive serum and requesting a new supply was not viable, Ashworth agreed to process the existing lot with the proviso that each batch would be subjected to triplicate safety testing.[104] Ashworth knew that failure to release a safe product would not only make Squibb vulnerable to litigation, but also risk the health of enrolled children and undermine the evaluation process. Ashworth mandated that following safety checks at Squibb, each batch would be shipped to the Plasma Fractionation Commission in Boston, Massachusetts, and to the Division of Biologics Control for additional testing.[105] Only when all organizations returned favorable results would the serums be released to Hammon. Even though triplicate safety testing was expensive, time consuming, and unprecedented, Ashworth reasoned that it was imperative to exceed regulatory standards, considering the prior handling of the GG and its forthcoming clinical application.[106]

Production delays came as a surprise to Hammon and NFIP officials, who were only beginning to appreciate the failures in central planning. One week before the pilot study was due to commence, Kumm contacted Squibb and was stunned to learn that the serums had not been bottled.[107] Like Hammon, Kumm assumed production would take only days and did not expect four to eight weeks to process, bottle, test, and package the serums. Continued uncertainty about serum production inspired frustration and blame. Ashworth attempted to allay fears and promised Kumm that serum bottling would commence on August 13 and be completed after two days. However, Hammon and NFIP officials neglected to account for the additional time required for triplicate safety testing. Squibb technicians required one week to test each batch followed by another full week assessment at the Division of Biologics Control and a subsequent two-day assessment at the Plasma Fractionation Commission.[108] Based on this projected schedule, the serums would not be cleared for release until the end of August.[109] Kumm telephoned Hammon, who was reportedly upset by the "disturbing news."[110] Since the potential efficacy of GG could only be measured during the summer months when severe polio outbreaks were most common, every day lost to safety testing meant less time to conduct the study. Prior misunderstandings and concern had pushed the start date to late in the polio season.

After experiencing disappointments and production deadlines, Hammon and NFIP officials devoted more attention to production oversight. Hammon asked the NFIP to inspect production, since its headquarters was located near Squibb's plant. To improve communication, Kumm visited the Squibb plant on August 14 to witness serum bottling. His fears were placated during the tour, and he later assured Hammon that production appeared to be proceeding on schedule. "Dr. Ashworth showed me the gamma globulin and gelatin solutions. They look almost identical," he wrote. "Unless something unforeseen happens, therefore, the gamma globulin and gelatin should be ready before the end of next week."[111]

Despite improvements in production oversight and coordination, the legacy of project management failures was difficult to overcome. Soon after Squibb personnel began safety testing, contamination in one of the GG batches resurrected fears. On August 21, Squibb laboratory technicians discovered that one vial in fifty samples of GG "showed a gram positive coccus," suggesting bacterial contamination. Ashworth was concerned, but reasoned it was likely due to a "mistake" at the Squibb laboratory and ordered a retest of ten samples.[112] Since Ashworth refused to send on samples to the Plasma Fractionation Commission until internal safety tests were successful, the release date for the serums was revised again. Although Hammon and NFIP officials had not anticipated production delays, they realized that the pilot study was doomed unless further intervention was undertaken.

Hammon used his clout and professional connections to expedite serum safety testing. He first wrote to Ashworth, expressing his disappointment at the delays and explaining that "the peak of epidemics for this year will soon be reached; and unless we can undertake our study prior to the peak in some areas, we will have to wait until next year." Ashworth appreciated the situation, but was unwilling to curtail his testing procedure. Hammon next lobbied the Division of Biologics Control, and in conversation with its acting director, he inquired "whether it would be possible to run sterility tests" concurrently with Squibb. Although Biologics Control rarely conducted parallel safety testing alongside manufacturers, Hammon persuaded the acting director that his study was a special case where such considerations were an absolute "necessity for saving time."[113] With the assent of the Division of Biologics Control, Hammon obtained an unusual procedural exception.

To further compensate for lost time, some NFIP officials were sent to Squibb to help its personnel randomize, record, and package the nearly five thousand bottled serums. Since the placebo and GG solutions needed to arrive at the proving ground visibly identical yet randomly distributed, a randomized packaging process was developed.[114] Drawing on a time-honored randomization method, Kumm assigned the distribution of GG and gelatin solutions using white and black marbles.[115] Each bottle was numbered and packaged according to the random selection of marbles and the key to the sequence was recorded and locked away at NFIP headquarters for Hammon's later reference.[116] The randomization process demanded precision so that the test serums were accurately allocated and documented. Although too late to assure a mid-August release date for the test serums, the subsequent intervention and remedial action orchestrated by Hammon and NFIP officials salvaged the most critical component of their pending scientific experiment.

While the production process moved ahead, Hammon turned to the challenging task of selecting a suitable civilian proving ground. Even though polio outbreaks were notoriously difficult to predict, as well as widely variable in magnitude and duration, trial administrators sought to transcend uncertainty and locate the perfect epidemic.[117] Since quickly identifying an outbreak before the end of the polio season was imperative to the success of the experiment, trial administrators invested in a novel forecasting system that harnessed epidemiological surveillance, statistical modeling, and local assessments. Gathering and tabulating data on emerging outbreaks was the first step. Hammon reasoned that only through the accumulation of timely polio incidence data from each state could he pinpoint an early-stage outbreak of suitable proportions. To centralize reporting, NFIP personnel expanded their epidemiological surveillance network by requesting that states send "weekly telegraphic reports of poliomyelitis morbidity [incidence of the disease]" to headquarters.[118] This program soon rivalled the federal government's polio epidemic monitoring program managed by the Communicable Disease Center in Atlanta, Georgia. Although state health officers were not required to comply, the NFIP had nurtured strong working relationships with most health departments, and many cooperated. Under the supervision of Dr. Harry Weaver, the NFIP created an epidemic monitoring unit that tabulated incoming polio incidence data from sixty counties and prepared it for trend analysis.[119]

The intensive monitoring system was complemented by a system to predict the potential intensity and longevity of a given outbreak. The practice of using statistics and probability was rooted in World War II, when military strategists employed mathematicians to reduce risks and increase victories.[120] To achieve a similar feat of forecasting, the NFIP turned to University of Michigan statistician Dr. Fay M. Hemphill.[121] Hemphill was experienced using medical statistics and published theories ranging from polio epidemic recurrence to models predicting the "expected epidemic period" of an outbreak.[122] It was this latter concept that roused the interest of the NFIP. Hemphill believed that it was possible to compare current polio incidence data with historical data and locate "distributions of recognizable similarity for the years of greatest total cases"; in other words, he claimed to have a means for determining the intensity and duration of an epidemic.[123] Although Hemphill did not promise definitive results, his data helped reduce the perceived randomness of choosing an epidemic.

As well as personnel for epidemic monitoring and statistical prediction, a surveillance team was assembled and deployed. Hammon and NFIP officials believed it was necessary to have experienced personnel assess local facilities, community cohesion, and the nature of the outbreak before the test site was finally chosen.[124] Traveling incognito under the auspices of an epidemiological study, trial administrators scouted several communities. NFIP assistant medical research director Dr. Henry Kumm scrutinized Fort Worth, Texas, and Colorado Springs, Colorado, while field trial deputy administrator Coriell evaluated the suitability of Zanesville, Ohio.[125] Hammon, meanwhile, assessed several townships in Illinois and a site in Utah County, Utah.[126] Personal assessments enabled researchers to weigh selection criteria and float the idea of a pilot study to local doctors and public health officials before a formal commitment was solicited. Unbeknownst to residents, Hammon and his team were moving closer to identifying a civilian test site.

Even though epidemic monitoring and statistical predictions refined the options, Hammon ultimately chose the proving ground based on firsthand observations.[127] The small towns of Colorado Springs and Pueblo, Colorado, were initially strong candidates, but by late August there were "indications that the [epidemic] peak" had been reached.[128] By contrast, the outbreak in Utah County, Utah, was at an early stage with evidence of mounting severity.[129]

Health officers had reported seventeen cases of polio in the last week of August and seven cases the week prior, as well as a high incidence of bulbar polio and a 10 percent death rate.[130] Utah County was also favorable for cultural and political reasons. Encompassing miles of countryside bordered by the Rocky Mountains and replete with fruit farms and small towns, such as Springville, Spanish Fork, and Orem, Utah County boasted a cohesive society. Provo, the county capital, had a population of twenty-nine thousand and was situated forty-five miles south of the state capital, Salt Lake City.[131] The county was largely homogeneous in terms of race and ethnicity. According to the 1950 census, ninety-nine percent of Utah residents were white.[132] A shared religion added solidarity, as the state was settled by followers of the Church of Jesus Christ of Latter Day Saints (LDS) and some 70 percent of county residents were affiliated. Utah County was also a convenient location, since Hammon had selected Salt Lake City as a staging area for personnel and equipment. With time running out on the 1951 polio season, Hammon and NFIP officials gambled on Utah County.

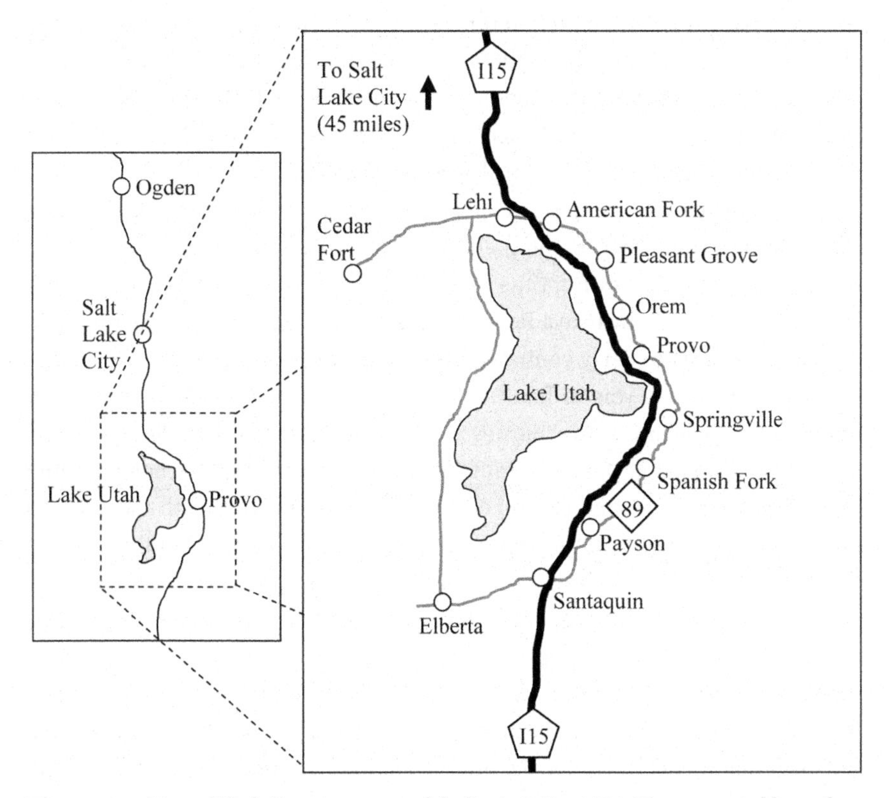

Figure 10 Map of Utah County gamma globulin test site, 1951. Figure created by author.

"Here we come and God Save the King," announced NFIP radio director John Becker before departing New York City on August 30, 1951, for Salt Lake City, Utah.[133] Hammon and the NFIP rushed to mobilize equipment, supplies, and personnel. Staff, in particular, needed to be trained and injection clinics established and supplied—all within mere days. Hammon understood that a failure in Utah County would be costly, not only in financial terms, but also in assuring the continuation of the experiment. While the publicity campaign was primed for release, new logistical deficiencies threatened to delay the study.

Transporting equipment and preparing supplies was challenging under the time constraints. NFIP assistant research director Dr. Henry Kumm picked up the test serums from Squibb, but was nearly delayed at LaGuardia Airport when the flight could not accommodate the extra 445 pounds of supplies.[134] Quick-thinking airline staff and accommodating passengers came to the rescue. Kumm later reflected that "five passengers were asked to give up their seats to permit the gamma globulin to travel on the overloaded plane."[135] Moreover, the needles, syringes, and other necessary equipment were shipped by train from the University of Pittsburgh to Salt Lake City.[136] While awaiting their arrival, Hammon realized that the limited infrastructure in Utah County would complicate the field study; arrangements for the sterilization of needles and syringes were hastily agreed on with one Salt Lake City hospital; Hammon negotiated sterilization in several nightly batches.[137] The limited supply of equipment allowed little room for delays, breakages, or accidents.

Time constraints also posed challenges in transporting, lodging, and training assistant clinic staff. To help administer the clinics, the NFIP contracted twenty-five polio nurses from the Joint Orthopedic Nursing Advisory Service (JONAS) and six medical residents. NFIP officials originally planned to charter a private flight to carry contract personnel to the test site, but the plan was scrapped because it was difficult to organize in time.[138] Instead, contractors organized their own railway journeys to Salt Lake City, while lodging arrangements were booked at the last minute.[139] It was not until the day before clinics opened that Hammon and Coriell convened a joint meeting with the contract medical and nursing staff to discuss "the technique to be used in the inoculation centers."[140] The medical residents were to conduct injections under the supervision of Utah doctors, while the nurses were to coordinate the clinics, assist with injections, manage records, supervise volunteers, and adapt to unexpected circumstances.[141] Although the consultants were well-qualified for their duties, the hurried training likely posed difficulties in learning about the experiment and its protocols. Since the proving ground was selected with little prior knowledge, complex staffing issues could only be addressed as they became evident.

Despite numerous logistical failures and the absence of a project manager, Hammon and the NFIP worked together to resolve emergent problems and establish the operational base for their clinical trial. Through this period of transition, the power and resources of the NFIP were mobilized to push the experiment onward despite uncertainty and shortcomings. With the clinics and local medical alliances in place, the moment had come to announce the GG pilot study to the parents of Utah County.

Chapter 4

The Pilot Study

At a press conference announcing the gamma globulin (GG) pilot study in Utah County, four-year-old Kristine Hammond was volunteered by her physician-father, Dr. Roy Hammond, to be the first child recipient of the test serums. Kristine's photograph, showing her wearing a crisp, pleated dress, with her hair in curls and looking up into the face of Dr. William McD. Hammon, sent a clear message that the pilot study was an important event overseen by responsible researchers.[1] The *Provo Daily Herald* ran a feature showing Roy Hammond seated with his daughter, implying both parental and professional approval.[2] These buoyant visual characterizations were part of a closely managed public relations campaign implemented by the National Foundation for Infantile Paralysis (NFIP) and Hammon to influence perceptions of the impending experiment. Although considerable attention was devoted to public relations, no one at the time knew whether parents would step forward to volunteer their healthy children to the study. How did members of the community react to the public relations campaign? What did scientists learn from their activities? This chapter explores how Hammon and the NFIP approached publicity, the tactics they deployed, and the public response. It also examines how thousands of children were mobilized to assess the safety and effectiveness of GG.

Seeking Approval of Government Officials

Once Hammon identified Utah County for his GG pilot study, he acted quickly to set in motion the marketing and mobilization plans by contacting high-ranking Utah bureaucrats and health professionals. On August 30, 1951, he met with senior members of the Utah State Health Department, who granted "clearance" to proceed. He next consulted with the Salt Lake County and Utah State medical societies.[3] During these meetings, Hammon mentioned the prior sanction of state health officers, as well as that of the NFIP Committee on Immunization, to obtain "approval to go ahead." State medical society officials were especially enthusiastic about the study and offered to "obtain additional physicians for the project in Provo should they be necessary."[4] With senior doctors and health officers on board with the study, Hammon believed he was prepared to solicit the Utah County medical society.

Hammon traveled to the county capital, Provo, where he met privately with representatives of the medical society "so that no embarrassment could come to the medical profession should they believe it unwise to endorse and participate in the project."[5] He discussed the nature of the study with society president Dr. Roy B. Hammond and society secretary Dr. R. H. Wakefield, who both agreed to "an emergency meeting of the total membership" so that Hammon could deliver a "carefully prepared address describing the plan for the field project."[6] Thirty-six society members were invited, as were county and district public health officers.[7] Those unable to attend were "contacted by telegram or telephone."[8] The excitement generated by the emergency gathering provided Hammon with the perfect venue to sell the pilot study.

The address to the Utah County medical society was a masterpiece in persuasive marketing.[9] Hammon impressed on county doctors that the GG study was part of a rational scientific trajectory. "Our experiment is a logical sequence to the previous experiments, and designed to correct the deficiencies of the others," he enthusiastically stated.[10] A rigorous assessment of GG for the prevention of polio was not an aberration, but a process linked to the scientific method. He then invited doctors to become partners in science by promising that the "clinics will be your clinics and we wish to place our small group of pediatric residents, who are available to assist in inoculation, under your direct supervision."[11] By assuring county doctors control over the study, he cast it as a grassroots venture; injection clinics were not zones of northeastern medical imperialism, but places of civic activism.[12] Members of Hammon's field team were situated as consultants, providing the necessary resources and advice to realize an opportunity to protect the community. He further explained that participation in the study was optional and that the decision to opt-in lay with Utah doctors. Respecting the authority of physicians, Hammon stated, "We propose to inform you as completely as possible in respect to our plans and purposes, and to withhold nothing whatsoever. . . . If our plan does not meet with your approval, we have no intention of carrying it out."[13] Hammon's pretension of openness and deference to clinical judgment placed his audience at ease.

Complementing the deferential tone, Hammon offered county doctors special aid and resources. He volunteered his field team to help doctors undertake "follow-up" work after the injection clinics were closed. He explained that each emergent case of paralytic polio, once reported to the health department, would be seen by his chief clinical consultant to standardize diagnosis and reporting. He assured physicians that the information gathered would be used "for assistance in treating the case." The willingness of Hammon to share data portrayed a commitment to the local community and an investment in the results. He assured physicians that his follow-up team would not interfere with "care or treatment," and would be available to consult if "requested by one of you."[14] Physicians would be supported in their battle against polio and by participating in the experiment they would enjoy the expertise and resources of leading researchers. For many doctors, likely overwhelmed by the outbreak, the promise of assistance with diagnosis and treatment was a welcome relief.

While Hammon's address did much to assure physicians, it omitted and minimized potential problems. Scientific debates about the study's methodological shortcomings or potential health risks were either marginalized or ignored. The address anticipated the possible reticence of physicians and the legitimacy of their fears. "Some of you," he began, "I am sure, are concerned about injecting inert, control material to half the volunteers during the epidemic . . . because the injections might induce paralysis." Hammon explained that since injected gelatin did not contain antigens it would not endanger healthy children. He also sidestepped the possible dangers of serum hepatitis by professing that donor blood was safe once fractionated.[15] Crowd contagion was altogether ignored in the address. By sidelining potential health risks, Hammon presented the pilot study as no more worrisome than any other public health program.

Utah County doctors sought participation in Hammon's study for personal and professional reasons. Their desperation during a moment of crisis proved to be an important catalyst. At the time of the meeting, twenty polio cases had been reported, with new cases emerging daily.[16] Endorsing the study was, for many doctors, a rational decision based on the notion that action was preferable to inaction.[17] The sense of urgency and the powerlessness of physicians in the face of the epidemic stifled debate and incited cooperation. Concerns about family members added additional pressure to physician-fathers. As shifting concepts of masculinity during the Cold War increasingly normalized men's domestic roles as consumers and parents, investment in children increased male concern for their well-being.[18] By supporting the experiment, some physician-fathers empowered themselves as parents and protectors of the family.

The opportunity to participate in the creation of scientific knowledge was appealing to local physicians, as the demands of family practice had prevented most from contributing to formal research.[19] It was likely that many were flattered by the proposed assistance and eager to be part of a possible breakthrough in the prevention of polio. Furthermore, some physicians likely imagined that participation would bring them recognition as medical innovators.[20] Participation in the study was not only a chance for professional fulfillment, but also a public relations opportunity.

The dynamics of the medical society likely shaped perceptions. As a relatively tight-knit group of male professionals sharing a religious affiliation, Utah County physicians were perhaps susceptible to group consensus. Within this hierarchy, the opinion of society president Dr. Roy Hammond held considerable weight. Born in Salt Lake City, Hammond attended Brigham Young University and subsequently graduated with honors in 1940 from the George Washington Medical School in Washington, D.C.[21] Marriage to Anita Smoot, daughter of Republican senator and church elder Reed Smoot, united him to a political dynasty and the Latter Day Saints (LDS) aristocracy.[22] This locally prominent lineage assured Hammond unique influence in the community.

As Hammond was among the first members of the county medical society to meet Hammon, he had the opportunity to reach an opinion of the study before the address began.[23] Dr. Lewis L. Coriell's prior experience in the biological warfare service, Hammon's war consultancy in the Pacific, Dr. Francis Sargent Cheever's epidemiological service in the navy, and Dr. Roy Hammond's role as a flight surgeon in New Guinea perhaps drew them together as allies facing a common enemy.[24] Their shared gender, class, ethnicity, war experience, and Cold War faith in science probably fostered a measure of trust and fraternal solidarity that transcended the formality and technical specifications of the protocol.[25] Even though Hammond did not blindly endorse the experiment, evidence suggests that first impressions mattered and his positive opinion affected the attitudes of fellow clinicians.[26] When the medical society vote was called, personal, professional, and political realities converged. Physicians reportedly "decided unanimously to support" the pilot study with "unexpected enthusiasm."[27]

After obtaining the approval of doctors, Hammon shifted attention to civic and school officials. On August 31, he met with C. M. Aldrich, acting mayor of Provo, who gave his approval of the experiment and permitted use of public buildings.[28] Hammon mentioned he was "satisfied" with Aldrich's belief that residents would want to participate in the pilot study.[29] Hammon continued his solicitations by approaching school superintendent J. C. Mofatt.[30] Like other administrators, Mofatt "heartily endorsed" the study and granted the use of school facilities to house injection clinics.[31] The support of civic, county, and school officials assured that local infrastructure would be available for the study. With the backing of health professionals and civic leaders, Hammon was ready to announce the experiment and release the planned publicity program.

Winning the Support of Journalists and Broadcasters

A core strategy of the publicity program centered on the enlistment of state and local media outlets. Utah radio producers and station owners were approached to broadcast special announcements and interviews about the experiment, since radio was a powerful means of reaching much of the population.[32] KUTA radio in Salt Lake City interviewed Hammon and Coriell, which provided them with valuable airtime on a station followed by many Utah residents.[33] Subsequently, Utah County's KOVO radio in Provo held a noontime interview. Drs. Hammon and Coriell, and the NFIP assistant director of research, Dr. Henry Kumm, were joined by county physicians Drs. Hammond and Wakefield as well as by the state public health officer.[34] The county doctors and visiting scientists worked together to generate interest. Kumm noted that "the [radio] announcer asked a number of questions of the doctors," whose relaxed manner, comforting answers, and optimism helped reduce public anxieties.[35] The alliance with the media was off to a positive start.

Complementing radio broadcasts, Hammon and NFIP officials sought the assistance of newspaper journalists and editors. On August 31, NFIP officials convened a special press conference in Salt Lake City, where they invited representatives of the *Provo Daily Herald, Salt Lake Telegram, Deseret News*, the United Press, and the Associated Press. Hammon read from a script and fielded questions. Although some editors and journalists were concerned that insufficient time had been allocated "to condition the people," they nevertheless granted their support through front-page coverage, editorials, and human interest stories.[36] Kumm happily observed that "as a result of the meeting, the press extended excellent coverage to the entire program."[37] The front page of the *Salt Lake Telegram* announced "Utah County Chosen for First Mass Tests against Polio," with an article claiming that this was "one of the most significant experiments ever conducted."[38] Leo N. Perry, bureau news chief for the *Deseret News*, likewise wrote a flattering piece, emphasizing the novelty of the study and explaining to readers that "the March of Dimes project" was the "first of its kind on human beings."[39] Initial media coverage helped Hammon rouse public interest in the study before its launch.

Figure 11 Promoting the gamma globulin pilot study on KOVO Radio, 1951. Pictured (left to right) are Dr. Roy Hammond, Dr. Lewis L. Coriell, Dr. Gordon Johnson, Dr. Henry Kumm, Dr. William McD. Hammon, Dr. Harold Wakefield, and Dr. John Rupper. Courtesy of the March of Dimes Foundation.

A second media event was held on September 3, 1951, the day before the GG experiment commenced, where Hammon invited Utah reporters to a "dress rehearsal" of the injection clinics. With the assistance of local doctors, Hammon and Coriell submitted to injections of the test serums. One journalist explained that the scientists were among the "first to get 'shot.'" Hammon and Coriell's decision to use their own bodies to demonstrate faith in the study echoed similar practices among nineteenth-century medical researchers, "linking professionalism to personal honor and character."[40] Although the expectation that researchers would submit their bodies was waning by mid-twentieth century, its practice was part of a larger "moral tradition."[41] By submitting to the injections, Hammon and Coriell attempted to connect to this moral trope and uphold public perceptions of medical science as a responsible discipline occupied by concerned experts. The demonstration exhibited researcher's confidence, reinforced earlier media attention, and added hype surrounding the launch.

Following on from Hammon's demonstration, Utah County medical society president Dr. Roy Hammond submitted his four-year-old daughter, Kristine, as the first inoculated child.[42] Similarly, society secretary Dr. R. H. Wakefield volunteered his seven-year-old daughter, Lorraine, to be the first inoculated child when the clinics opened.[43] Like self-experimentation, enlisting close relatives to uphold professional convictions drew on historical precedents.[44] Since most Americans considered the parent-child bond to be sacred, the volunteering of offspring to an experiment was understood to be an act of faith. This sentiment was likely amplified within the LDS family-centric culture, which was protective of children.[45] As the family patriarch and guardian of his child, Hammond's actions sent a strong message to the LDS community that the study was safe for children.

Utah journalists and broadcasters also helped normalize the form of consent, previously developed by the NFIP's legal consultant and "printed in local newspapers and read over the radio so that parents could be fully informed and understand what they would sign before coming to the clinic."[46] Media coverage of the form provided parents with time to consider its implications and reach a consensus before seeking participation. Hammon hoped that the inclusion of the form in newspapers would streamline clinic enrollment, since "parents could study it and might even clip it out to sign and send with older children."[47] The *Daily Herald* reproduced the form and explained that "any parent who might send his child alone to the clinic testing station must send a signed consent form with him or her."[48] Since media coverage of the form did not address health risks, Utah newspapers situated the legal waiver as a reasonable expectation and an uncontroversial concept when seeking children's inclusion in a medical experiment.

Encouraged by the enthusiasm of Utah journalists, health professionals, and civic administrators, Hammon turned his attention to encouraging parents to participate in the study. Although he had not anticipated the need to mobilize religious organizations, he quickly appreciated their importance, as Utah County was an LDS stronghold.[49] Hammon used his connections with the local medical society to access the highest echelons of church bureaucracy. Winning over the stake presidents, who administered religious districts and held authority over church mandates, was imperative.[50] Since most Utah County doctors held membership in the church, their opinions influenced the perceptions of presiding stake presidents within the institutional hierarchy.[51] On September 1, Hammon contacted stake presidents in four districts and invited them to support the study. Based on the earlier sanction of the medical society, the stake presidents unanimously agreed to endorse the project and help build support within congregations.[52]

The stake presidents released a special announcement in the *Deseret News*, urging "Church members to participate" in the GG study.[53] The *Daily Herald* echoed that stake presidents "urged its whole-hearted support by parents of children within the desired age limits." They reiterated that "the only way to test this serum, in the opinion of the best medical authorities, is in the manner now underway, and people of this area should do everything possible to insure its success."[54] The announcement invoked themes of medical expertise, trust, and accountability.[55] "The best medical authorities declare there is nothing harmful in the inoculations that will be given," they assured. Participation was further encouraged on humanitarian grounds "for the benefit of children all over the world."[56] The stake presidents' announcement positioned the GG study as an exceptional opportunity and a reason for hope during a public health emergency.

Complementing the press release, stake presidents permitted Hammon to prepare an announcement for delivery at Sunday church services. The resulting script attempted to captivate lay audiences and allay their fears. The outcome of animal experiments was described, as well as the outcome of earlier GG studies in humans. The safety of the test serums was backed by the promise that "there was no evidence that either substance could possibly be harmful." Listeners were assured that the selflessness of volunteerism would benefit all nations and peoples. "Only a good turn-out at the clinics," the script noted, "will make it possible for doctors all over the world to know whether this serum can be used successfully to protect children against paralysis and death, during future epidemics."[57] Although the international implications of GG for polio had not been previously envisioned, the rhetoric echoed with the LDS missionary program.[58] As a former minister and Methodist missionary, Hammon understood the power of linking the shared humanitarian mandates of religious and scientific pursuits; he also knew that delivery of this script through a credible platform would maximize awareness and support. When Hammon's script was read aloud to congregations on Sunday, September 2, popular support was galvanized.[59] Since worshippers were expected to heed the endorsement with the same ardor shown toward other official mandates, the community was primed for participation.

While parents were influenced by doctors and LDS leaders, they also chose to participate for their own reasons. Most residents were frightened by the polio epidemic and optimistic about the promise of a new medical intervention. On

September 7, seventeen-year-old Jean LaRae Pulley from Pleasant Grove succumbed to bulbar polio at the Utah Valley Hospital. Active in the LDS church and a participant in various music and dramatic performances, Pulley was the fifth polio death in Utah County that year.[60] After witnessing the death of his daughter, Ray Pulley enrolled his younger children David and Lynn in the GG study. According to observers, the Pulleys "were among the first in line."[61] Like Pulley, Mrs. Elaine (Claude) Robbins's firsthand experiences with the consequences of polio inspired her support of the study.[62] The epidemic stuck two of Robbins's four children: Carl Claude Jr., age seven, suffered bulbar polio, while Katheryn, age nine, experienced paralysis of the extremities. At the time of the newspaper interview, doctors had discharged Carl to home care, while Katheryn remained in the Salt Lake City hospital. The *Provo Daily Herald* used Elaine Robbins's testimony to weave a timely story of endorsement. "Anything that anyone can do to save other children from the terrors of polio certainly should be done," she explained. Invoking notions of duty and concern for the welfare of others, Robbins continued that "if this experiment can do anything to keep children from suffering as she [Katheryn] has suffered, we would be inhuman not to do everything we can to help it out."[63] To demonstrate her resolve, Robbins intended to enroll her healthy four-year-old daughter, Mary Ann, in Hammon's study. The larger crisis for Robbins was the devastation caused by polio, its probability of affecting other family members, and her powerlessness to prevent it.

Like parents, many children were afraid of polio. As former study participant James Jex recalled, "We were always scared that we'd get polio if we got a scratch from something rusty. I had a friend later on that did have polio when he was a kid and he had one side that was involved in paralysis, but he made do the best he could with his situation."[64] Participation in the study offered families a sense of empowerment at a moment of helplessness.

Many parents trusted the opinions of local doctors. After the study was announced, doctors' offices were reportedly "flooded with phone calls for advice to inquiring parents, and apparently all [doctors] endorsed the program."[65] Drs. Hammond and Wakefield were not the only physicians to volunteer family members to the study. The *Deseret News* reported that "Utah County physicians set an example for the community by having their own children among the first at the clinics."[66] As respected professionals, county doctors' willingness to enroll their own children sent a strong message to the community that the experiment was safe and potentially useful.

The belief that participating in the study was less dangerous than waiting out the epidemic influenced many parents' decisions. This reasoning, known as "lesser harms," had roots in eighteenth-century medical practice to justify the "use of an intervention . . . if the risks it entails are lower than the risks of the natural disease that the intervention is designed to prevent."[67] As this ideology remained entrenched in American society, and since paralytic polio was already affecting dozens of children in the county, the lesser harm weighed heavily in favor of participation. Since parents were not provided with a full account of the health risks associated with the study, the net benefit of participation appeared self-evident.[68]

Volunteering children to advance medical knowledge was an important motivator for many parents. After enrolling his son in the study, Charles Loris of Provo explained to journalists, "This is the best thing to come to Utah. . . . It is something in which we can help the children of the future—not only mine, but everybody's in the world."[69] Discouraged by hitherto sluggish scientific advancements to control polio, some parents rationalized participation as a duty. One Utah mother explained that "because she had always blamed scientists for not making more progress," she chose to volunteer her child "to help them accomplish something definite."[70] Enrolling her child to help science was part of being a responsible modern parent.[71] The opportunity of becoming part of a medical discovery became a means for some families to fulfill a social obligation.

The nature of Cold War public health campaigns normalized participation in disease prevention programs. Because of the perceived threat of Soviet nuclear, biological, and chemical warfare, Utah was the site of civil defense initiatives between 1950 and 1951.[72] Under the direction of the State Defense Organization, health officers urged citizens to determine family members' blood types and immunize against smallpox, typhoid, diphtheria, and tetanus.[73] Complementing this agenda, the Utah Board of Education and Parent-Teacher Association (PTA) organized school immunization programs.[74] Health officers promoted the concept as a "good sense" measure and part of a mother's duty to protect her family. "True, a war scare means fear of bombing," one health officer explained, "and fear of bombing means fear of disease, but it's just good homemaking to get children immunized."[75] PTA volunteers were enrolled into "indoctrination" classes that stressed the importance of immunization.[76] Even before Hammon arrived in Utah, parents were accustomed to activist public health programs.

Some citizens not only sought to enroll their children, but also offered their assistance at the clinics. Women with the Utah County medical society's auxiliary division volunteered as clinic registration clerks, ushers, and aides.[77] A "house-to-house canvassing" campaign was also coordinated by local women's groups, but was soon called off because the popular response made it unnecessary.[78]

Although most parents supported the study, a few chose not to participate. Hammon encountered two forms of opposition: an "anti-guinea pig" attitude, as well as concern about crowd contagion.[79] NFIP officials noted that Hammon was "fumbling around for two days finding out what [was] wrong" and why some parents did not want to participate.[80] Reserve toward the study was not unusual, but it reflected a minority opinion.[81] For families already concerned by immunization health risks, participation in GG study held little appeal.

Hammon and his field team opened injection clinics at 9:00 A.M. on September 4, 1951. Clinics were established in Orem at the city hall, as well as in Provo at the Timpanogos School, Farrer Junior High School, Wymount Village Health Center, and the Franklin Elementary School. As clinics began to register enrollment, "long lines of children accompanied by fathers or mothers, awaited their turn in the hallways."[82] Additional clinics were soon established in Lehi, Spanish Fork, American Fork, and Pleasant Grove. Like their counterparts in Provo and Orem, these smaller clinics were soon overwhelmed. Although most families expressed frustration at the long lines and congested clinics, they remained undeterred.[83] Families attending the Spanish Fork clinic negotiated crisscrossing stairwells while they waited in the line.[84] Anxious children leaned on railings, straddled chairs, or held on to their parents for comfort. The public relations program seemed to have succeeded: "The final unknown element—the willingness of people to cooperate—was answered today when 1,459 children between the ages of two and eight were inoculated in an unprecedented human test," one NFIP release boasted.[85] Hammon likewise gushed, "It appears scientists may have underestimated [the] willingness of the public to participate in research projects."[86] The favorable public response pleased the scientists and helped the experiment gather momentum.

The injection clinics were central to the systematic enrollment of children. The clinic design incorporated the trappings of clinical medicine and a method to process hundreds of families per day.[87] Establishing injection clinics in public buildings, such as schools, health centers, and civic structures, made them visible and accessible to parents; trial administrators ensured that even small towns, such as Spanish Fork, with a population of five thousand residents, had at least one clinic.[88] The utilization of public venues upheld the significance of the study and the backing of civic and state officials. The fusion of bureaucratic sanction with scientific interests brought the concept of public participation in the experiment in line with existing public health programs and notions of civic responsibility.

Figure 12 Injection clinic at Spanish Fork, Utah, 1951. Courtesy of the March of Dimes Foundation.

Although many parents and their young children waited for the injections, the clinic design reinforced commitment and order. Most clinic lines were overseen by a "uniformed fireman" or police officer, who provided surveillance and assured proper conduct.[89] Line overseers were required to "maintain order in the waiting line," as well as "prohibit grouping, the exchange of personal articles, and horseplay."[90] In accordance with Cold War notions of conformity, line monitoring necessitated that families follow behavioral norms.[91] While surveillance achieved practical benefits, including the reduction of line-hopping or jostling, it also fostered a mild form of social control. This marked an important psychological and behavioral transition in which parents began to conform to the protocol and shepherd their children forward as research subjects.

The erosion of parental power occurred within the confines of the clinic. At the registration station, parents supplied personal information for each child and signed the form of consent in a setting managed by friends and neighbors.[92] Even though Hammon was initially concerned that some parents might be apprehensive about the form, he was pleased that "there

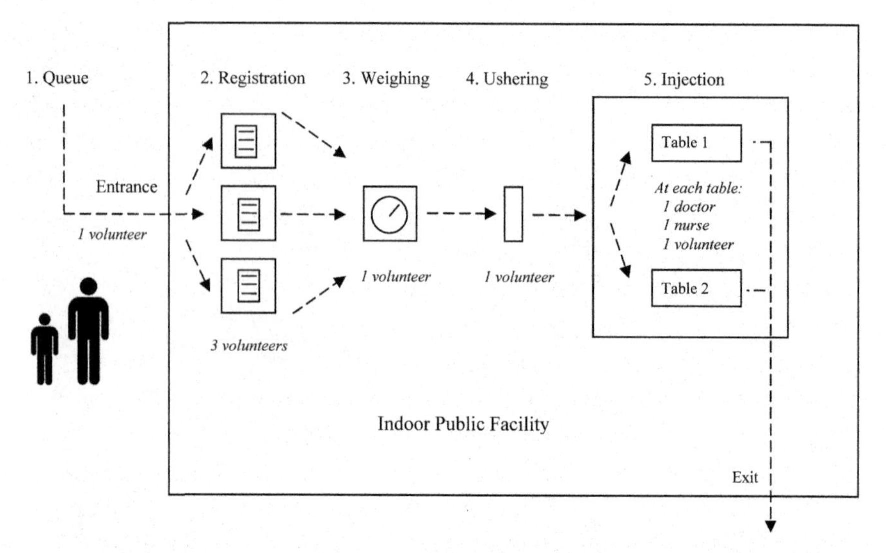

Figure 13 This diagram shows the process developed by Dr. William Hammon to facilitate the enrollment of thousands of children at the gamma globulin study injection clinics. Based on William McD. Hammon and Robert Pierce, "Diagram: Station #5, Injection Room," June 6, 1952. Courtesy of the March of Dimes Foundation.

was no objection on the part of the parents to this procedure."[93] The trappings of scientific precision were evident at the weighing station. Since each injection dose was correlated to the mass of each child, clinic volunteers weighed children prior to escorting them to a holding area on the fringes of the injection tables.[94] The injection station, as the final stage in the process, consolidated the transformation of the child to human subject. Two immunization tables were arranged to facilitate parallel injection teams.[95] Inasmuch as the setting resembled a triage center, it also contained reminders of scientific medicine: sanitation, order, and efficiency. Although physicians administered the injections, the traditional doctor-patient relationship was altered to permit efficient processing.[96]

Most parents were not discouraged by the 50 percent chance that their child would receive placebo. Hammon was surprised by this attitude and reported that placebos had "not fazed parents a bit." He reasoned that "most of them have taken the view that the two-fold goal of aiding science, while taking a fifty-fifty chance of perhaps protecting their children, [was] sufficiently rewarding."[97] NFIP officials likewise observed that "those questioned said the chance of getting the possibly protective serum was well worth the trip to the clinic."[98] When Charles Loris was asked whether he hoped his son would receive a dose of GG rather than gelatin solution, he replied, "It makes

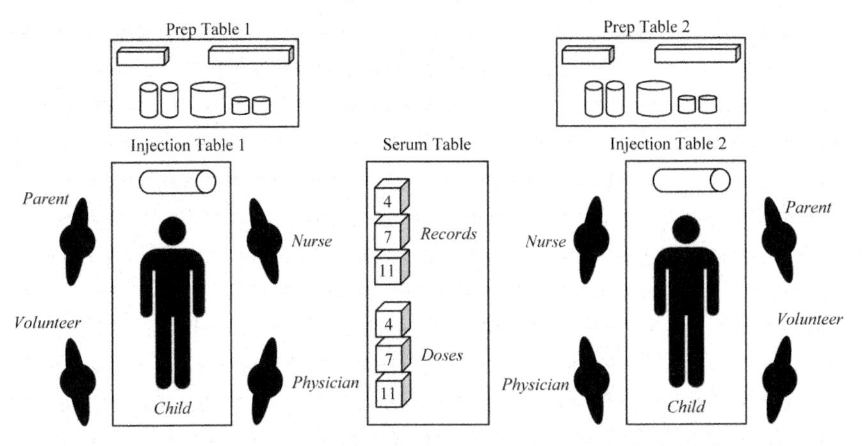

Figure 14 This diagram shows Dr. William Hammon's design for clinic injection station. Based on William McD. Hammon and Robert Pierce, "Diagram: Station #5, Injection Room," June 6, 1952. Courtesy of the March of Dimes Foundation.

absolutely no difference to me. We want to help with the experiment. It means so much to everybody."[99] Like Loris, most parents accepted the concept of a controlled trial out of hope and a desire to advance knowledge.

A few audacious parents chose to increase the probability of their child receiving GG. Hammon discovered after the second day of the study that some citizens were enrolling children at more than one clinic to obtain additional inoculations. Although it is unknown how many multiple enrollments occurred, this practice highlighted a tension between parental agency and the statistical needs of trial administrators. To detect and prevent the practice, Hammon required doctors to wipe a small amount of picric acid, a common laboratory phenol reagent, on the buttock of each participating child.[100] When the reagent came in contact with skin, the resulting brown color served as a temporary indicator of participation. It was evident that a minority of parents understood the implications of the placebo very well and attempted to circumvent the protocol to maximize their child's chance of possible protection.

Some anxious parents refused to wait for a clinic to open in their neighborhood. Although Hammon allocated serums based on community population, the assignments became unbalanced when some parents drove to neighboring communities to enroll their children. Clinic volunteers in Provo experienced an influx of parents "coming in from outlying areas." Hammon compensated "a little by adding to the Provo allotment" and shifted serum supplies from the rural clinics.[101] The need to shuffle supplies showed the eagerness of some parents to be included in the experiment and the difficulty in managing the public response. Moreover, the enthusiasm of some citizens led to cases of unauthorized redistribution. Hammon discovered that American Fork volunteers illicitly reassigned "several boxes of materials from the Lehi stack which [by contrast] looked very high and much larger than theirs." With some dejection he noted that "Lehi was deprived and American Fork got more than it should have had."[102] Citizen intervention in the serum supply chain revealed the power some volunteers exercised over the study and their belief that it would confer some benefit to the local population.

Children had little or no input into the decision to participate in the study, which aligned with societal and legal norms of the time. "Dad just told us 'we are going to get a shot for polio,'" remembered James Jex. "That's how I learned about it. And he had made arrangements for some of my cousins to go too."[103] Hammon and NFIP officials depended on parental power to deliver a sufficient pediatric cohort. The injection of 4 to 11 cc into the gluteus maximus muscle using an 18 gauge (1.3 mm diameter) needle was unusual by pediatric standards. By contrast, most pediatric immunizations employed a finer 22 gauge (0.72 mm diameter) needle and only 1 to 2 cc of serum.[104] Since a large dose of serum administered with a large needle was acutely painful, parents played a vital role in managing their children's discomfort and anxiety.[105]

Most enrolled children demonstrated remarkable courage at the clinics. Young Albert Loris stated to reporters, "If it is all right with Dad, then it's O.K. with me. . . . I'm not a bit scared."[106] Similarly, James Jex recalled being frightened by his pending injection, but determined not to show his anxiety. "I remember I was scared to death because I didn't like shots," he explained, "but I didn't dare show it much, with my cousins and friends there." The sound of crying or evidence of pain did little to stoke confidence. As Jex continued, "One of my good friends walked out from getting the shot with his hand on his fanny; I remember that picture really vividly."[107] Others, such as three-year-old Sharon James, apprehensively watched the injection of her brother Stephen at the Orem clinic. A *Deseret News* reporter observed that she "stood at the door, twisting her 'big' brother's sweater in her tiny hands and her brow wrinkling into an anxious frown each time Stephen made an outcry."[108] Another journalist reported how children "noted with gloomy curiosity the reactions of their friends as the inoculations were given."[109] Many children likely developed coping strategies to appear stoic in the face of painful injections administered in the presence of friends and family.

Injected children demonstrated a range of emotions, from restraint to resistance. A photograph from Farrer Junior High School showed a child stretched on an injection table, covering his mouth and holding a concentrated stare during his injection.[110] Similarly, a reporter observed that a five-year-old girl "gulped hard as she was injected, then turned to the doctor and said, 'now I won't get polio, will I?'"[111] *Salt Lake Tribune* reporter William C. Patrick summarized a clinic scene this way: "Some of the children marched stoically into the operating rooms to receive their inoculations, and submitted to the injections without a whimper. Others protested loudly."[112] Journalist Leon Perry observed that "some [children] were dragged crying and with hands over their eyes, into the room where they received their inoculations." Many children were hauled kicking and wailing to the injection tables and held down by three or more adults. Young Max Smart, who was volunteered by father Neff Smart of Orem on his fifth birthday, "tearfully complained that the big needle was not quite his idea of a birthday present."[113] Although a few children bore their injection with reserve, most demonstrated physical and emotional discomfort.

The pain experienced by some young children was often disregarded by adults. As neither parents nor clinic personnel were being injected with the serums, there was no shared reference point for the severity of discomfort.[114] As NFIP staffer John Becker observed, "If adults were being punctured, this no doubt would scare off some folks. But the mothers are volunteering on behalf of their very reluctant small fry, who don't have much to say about it."[115] Although the assigned serum dose and needle gauge exceeded most adult immunizations, Hammon did not acknowledge the possibility of the study causing severe pain. "Good cooperation," he noted, "and no reports of any undesirable reactions, physical or psychological."[116] For Hammon, the pain that children experienced was temporary and a reasonable price to pay for the chance of possible protection against polio.

The provision of lollipops to child participants served an important psychological and public relations role.[117] Hammon and NFIP officials recognized that candies would help parents in drawing children to the clinic and stem cries of discontent after injection. Hammon noted that the "lollipops were effective in stopping the crying after injection as the injected children walked out of the clinic room door past the line of waiting children."[118] The lollipop established the child as a bona fide hero worthy of recognition; it also served in the process of reverse transformation: from a human subject back into child. The parting lollipop became a means of enticement, a tool of appeasement, and a symbol of merit.

As parental support for the experiment crested between September 5 and 7, serum shortages led to clinic closures.[119] The enrollment of children was

Figure 15 Boy leaves health center after injection, 1951. Courtesy of the March of Dimes Foundation.

"so heavy" in Provo and Spanish Fork that three clinics "were forced to close their doors ahead of schedule."[120] Most parents met with news of clinic closures with frustration and disbelief.[121] "The only discontent expressed," Hammon reflected, was "the fact that thousands were unable to participate because of exhaustion of the biological supplies." Hammon was disappointed by what he saw as a missed research opportunity, since he reasoned he could have injected at least another thousand children. "There were indications that 75 to 90 percent of the population had planned to participate and would have done so had supplies been adequate," he noted.[122] The parental response showed Hammon that over the course of a week, participation in the study had grown and that demand outstripped supply.

Although many families expressed disappointment when clinics closed ahead of schedule, participation made most citizens feel they had made important contributions to science and the nation. Reporters claimed the study had "placed Utah County on the forefront of the poliomyelitis stricken world stage."[123] The praise of scientists added to the sense of achievement; Hammon expressed that he and his team "were highly gratified by the excellent response" and that "the experiment may not only make medical history" but "provide the basis for new research projects calling for public support."[124] The fact that 5,768 children were enrolled instead of the intended five thousand children was framed as an example of Utah's unique culture. Hammon claimed that it was one of the "most highly successful public responses in medical research annals."[125] Through laudatory statements, most residents felt they had played a vital role in the fight against polio.[126]

Victories and Shortcomings

As the GG pilot study follow-up teams prepared their final reports and disbanded in October 1951, Hammon and his allies celebrated. The publicity campaign worked and most participating families had shown unexpected gratitude. Early observations further hinted that morbidity associated with paralytic polio "decreased abruptly during the week that clinics were held."[127] Even though it was too soon to know whether GG offered protection, initial signs were promising.

The study was also a sociological victory for Hammon and NFIP officials. It revealed that citizens would willingly participate in a large medical experiment. "One purpose of this experiment fell in the range of the social sciences," Hammon noted, "to determine whether [the] public . . . would cooperate adequately in this type of a study to render it feasible. The answer to this is a clear cut 'yes.'"[128] Legal waivers, placebo controls, health risks, injection pain, long lines, and the absence of therapeutic assurances did not deter most parents from enrolling their children. A Cold War faith in scientific innovation combined with effective publicity and a polio epidemic were potent catalysts for cooperation.[129] The GG study served as a precedent for mass public experimentation and would be referenced by NFIP officials when they later considered the prospect of a large trial of the first polio vaccine.

Hammon's experience in Utah reaffirmed the assumption that local doctors would cooperate with medical researchers.[130] Such alliances were not without precedent and were openly recognized in such long-running medical experiments as the Tuskegee syphilis study.[131] However, the scale of cooperation in Utah was extraordinary: not only had physicians volunteered their healthy children and administered the injections, but they had helped control parental access to private doses of GG. "Local sales of gamma globulin were minimal and many doctors reported that they had refused all requests to inject children with gamma globulin in their offices," Hammon reported.[132] Restricting access to GG was imperative, since unreported use would ruin the statistical rigor of the study. Obtaining the support and compliance of local doctors had made the experiment possible.

The experiment also strengthened the alliance between polio researchers and the NFIP. After 1951, the Committee on Immunization became more central in steering the direction of polio research. Leading researchers, including Drs. Jonas Salk and Albert Sabin, continued their close association with the NFIP and remained important contributors to its advisory committees. Investment in Hammon showed the scientific community that the NFIP would fund large experiments, as well as manage logistics and publicity. "The representatives of the National Foundation were most helpful, considerate and we would have been seriously handicapped without them," Hammon later admitted.[133] The study positioned the NFIP as a courageous sponsor of applied medical research in the fight against polio.

Careful media coverage of the study also assured the NFIP important recognition. Although most national press agencies limited reporting of the experiment to discourage premature public enthusiasm, a few select publications ran stories vetted by the NFIP.[134] A September edition of *Life* magazine ran a photograph-rich two-page article under the headline "Project Needle-Lollipop." The article created a flattering impression of the experiment as a hopeful venture, proclaiming, "Should it [GG] prove able to make the crippling effects less severe, it will be the best weapon against polio yet devised." The harried clinics, long lines, and painful experiences of children were not discussed.[135]

Despite evident successes, the GG pilot study suffered from serious deficiencies in promotion, design, and conduct. Even though Hammon professed to provide "full and factual information" and "to withhold nothing" about the study, the methods used to influence public opinion did not realize these aims.[136] Without a candid admission of the health risks associated with the study or the purpose behind the subsequent postcard survey, Hammon failed to provide adequate information to trusting parents to meet the standard of informed consent.

The study also exposed enrolled children to unfortunate clinical realities, as hurried injections increased the risk of injury and contamination. Although most doctors adhered to the protocol and followed cautious methods, some clinic observers noted inadvertent deviation. "Some of the nurses were a little horrified at the way some of the doctors handled the needles," one NFIP staffer described, "not keeping them as sterile as they might." He continued, "After all, when a doctor is used to giving inoculations in his office, he is quite careful but when it comes to an experiment, he may not be as careful. The doctors don't quite understand that the slightest contamination might tend to add another variable and change many of the answers that might be found from an experiment."[137] Although nurses were expected to offer advice to doctors, they had no authority to oversee the injections and were likely uncomfortable intervening to amend practices.[138]

Adverse health reactions were also observed among a few participating children. Although undocumented in 1951, injected GG could cause an allergic reaction in susceptible individuals.[139] During the study one child reportedly developed hives "on the right buttock, spreading to the leg a few hours after injection."[140] Although the rash coincided with injection, Hammon did not consider it a serious issue. Other children experienced a range of complex physiological reactions. A few children "fainted after inoculation," but Hammon dismissed such instances because they reportedly "had a history of having done so previously."[141] After the study, several children were taken to doctors "because of fever and malaise."[142] Although sources do not reveal how doctors treated these cases, none were hospitalized. Considering that local doctors' prestige and legal responsibility were tied to the study, there was little impetus for them to draw attention to adverse health reactions.

The study did not protect families from contagion or measure whether participation had exacerbated the epidemic. Though Hammon attempted to mitigate the dangers of viral exposure by instituting enhanced sterilization procedures and orderly lines, he found it difficult to adapt to local realities. In many cases, clinic facilities were unsuitable, since the rooms in public buildings were reportedly "much too small, especially the waiting room." Even though public health literature advised against grouping during outbreaks, clinic volunteers could not isolate participants or address the risk of contagion for fear of alienating parents. "The main objection I had to the medical aspect of this," John Becker recalled, "was the crowding of children together into a small room. Even though it might not do any damage, it puts the Foundation in a very bad light. We preach that children should avoid crowds during an epidemic and then turn around and actually put them into a crowd."[143] Like Becker, Hammon acknowledged that his experiment may have increased the spread of poliovirus in the community.[144] Although it was impossible to calculate the risk of contagion, the clinic environment did little to shield families from the epidemic.

The pilot study also failed to resolve the lingering uncertainty over serum hepatitis contamination. Months after the study ended, Hammon mailed the specially designed postcard survey to Utah parents, requesting information about emergent childhood illnesses, including cases of "yellow jaundice."[145]

While 91 percent of the postcards were returned, reporting accuracy was uncertain, and 523 reports were not returned.[146] Nevertheless, the survey revealed two cases of hepatitis in the cohort, which Hammon tracked with the assistance of local doctors.[147] Although placated by a subsequent investigation, Hammon's survey did not offer closure on the risk of serum hepatitis.

Concern about the risk of polio provocation also remained unanswered. Hammon prepared a confidential report, comparing the incidence of paralysis among injected and non-injected children, but admitted that conclusions were difficult to reach because his protocol was never designed to accurately track polio provocation.[148] "The actual number of cases occurring in the various groups [is] too small," he admitted, "to permit the drawing of statistically significant conclusions."[149] Despite a limited data set, Hammon did not believe he saw evidence that the injections caused polio paralysis.[150] Even though the study represented a compromise to assess possible harms, some health risks remained unknown.[151]

The quality of data collected during the experiment also fell short of expectations. Although the protocol followed controlled trial methodology, cases of clandestine serum allocation and multiple clinic enrollments ruined the data set.[152] Clerical errors further undermined accurate reporting. "In a number of instances," Hammon admitted, "I recall seeing that the doctor was confused and had recorded the number on the card of the wrong child."[153] Reporting of subsequent polio cases in the population was affected by personal judgments and ambiguous classification criteria. Although muscle examinations and spinal taps identified some cases of polio, a few patients were not classified. "We have three [cases] that could be classified as probably abortive [polio] among the injected who were in contact with a known patient," Hammon acknowledged. "Those were so mild that they were not hospitalized and no lumbar puncture was done."[154] Since the distinction between severe, mild, and abortive paralysis was not always easy to classify, not all cases could be recorded accurately.[155] Clinical assessments were also undermined by what children would tolerate. In one instance, "attempts to make a muscle evaluation . . . [were] unsatisfactory since the child [turned] out to be a fighting, kicking, screaming demon."[156] Although accurate reporting was central to the successful evaluation of GG, human agency shaped the data set in ways that could not be easily identified or corrected.

The outcome of the GG pilot study highlighted the challenges that trial administrators faced when conducting a medical experiment in an open population. Although Hammon and his allies achieved public relations victories and recorded sociological discoveries, the pilot study was plagued by adverse reactions, unknown health risks, and uncertain data quality. As presaged by some scientists on the Committee on Immunization, Hammon could neither eliminate risks nor assure the collection of objective data, but could merely test the boundaries of what the public would tolerate. Since it was evident that the public could be mobilized for science, Hammon stood poised to begin the next phase of his experiment.

Chapter 5

Operation Marbles and Lollipops

As a polio epidemic swept across Iowa in July 1952, Colonel John A. Carey, commander of the 79th U.S. Air Force Squadron based in Sioux City, learned of a field trial being conducted by Dr. William McD. Hammon to assess whether gamma globulin (GG) could curtail the epidemic and protect youngsters from paralysis. The squadron was not immune to the ravages of the outbreak; indeed, Airman Eldon O. Paul was admitted to nearby Saint Joseph Mercy Hospital with evidence of paralysis.[1] Although charged with preparing air crews for deployment in the fight against communism in Korea, Colonel Carey perhaps considered the fight against polio as an equally righteous and patriotic mission. In the midst of this local emergency, Carey assigned twenty-one squadron personnel to work at Hammon's injection clinics and speed science to an answer. Carey's group became an instant success with parents and journalists, leading the experiment to be dubbed "Operation Lollipop."[2] How did the next phase of Hammon's experiment become a reality? What were the results? This chapter assesses the expansion of the GG experiment in 1952, as well as how Hammon and the NFIP worked together to craft a story of scientific success.

Building Support for an Expanded Program

In the wake of the mixed success in Utah County, Hammon decided to prepare for an expansion of the GG study.[3] While 5,785 Utah children participated in the pilot phase, statisticians estimated that an additional fifty thousand children would be needed to achieve a definitive answer on the value of GG for preventing polio paralysis. NFIP officials reconvened the Committee on Immunization to discuss the outcome of Hammon's pilot study, sanction its continuation, and explore other research developments.

Hammon faced the committee on December 4, 1951, at the Hotel Commodore in New York City. NFIP research director Dr. Harry Weaver opened the meeting and asked Hammon to recount his experiences in Utah County. Hammon reported that the study was a success and that observed adverse reactions were of little concern. "In every reported instance," he explained, "the physician found some satisfactory explanation for the signs and symptoms, which appeared to be more likely to have been the cause than the injection." Hammon's rapport with Utah doctors probably kept negative reporting to a minimum. He presented the risks of polio provocation, serum hepatitis, and crowd contagion as either under investigation or impossible to measure. "Numbers are . . . too small," he explained, "to lead to any conclusions regarding possible increased incidence of disease or localization of paralysis."[4] The positive interactions between scientists, doctors, and the public were emphasized. "Interest and active cooperation were immediately obvious," he boasted. He claimed that thanks to the support of doctors and journalists, over 90 percent of Utah County parents wanted to enroll their children. He further asserted that the study, far from being a perilous venture, had ultimately won prestige. "The only criticisms heard," he noted, "were from persons who were turned away, and their complaints were frequently of a nature that caused us to regret deeply that we were limited to 5,000 ampoules for the study."[5] By portraying the study as a success, Hammon transformed it into a victory in the fight against polio.

The presentation justified continuation of the experiment by focusing on a need for more data and the promise of improvement. Hammon explained to committee members that the success of the overall program hinged on its extension. "We propose to continue the field project in 1952 along lines similar to those of 1951," he stated confidently. The data were presented as valuable and ready to combine with future results. "The [pilot] study is not a complete waste of time and money from this standpoint," Hammon assured members.[6] The next phase of the study, in addition to being undertaken on a larger scale, would be safer and more efficient. Instead of the cumbersome glass ampules, syringes, and needles, he promised that disposable syringes preloaded with serum would be used; this would not only reduce contamination risks, but would also increase the speed at which children could be processed. Hammon hoped that the need for more data and evidence of improvement in the protocol would satisfy committee members.

After his presentation, Hammon requested committee approval to continue the study. "Since plans are essentially the same as those already followed," Hammon stated, "I feel I can spare your time for other more difficult problems of the day, rather than to review this in more detail." Debate served no purpose for Hammon, who believed it was his decision to weigh the risks and benefits of the study. "Gentlemen," NFIP research director Dr. Harry Weaver stated, "you have a recommendation from Dr. Hammon for extending this study to the point of significance. What are your wishes?" As moved by Dr. Thomas Francis Jr., and seconded by Dr. Albert Sabin, delegates approved the extension.[7] Since the reported health issues appeared trivial, there was little reason to dissent. Through the consent of scientific researchers, Hammon and NFIP officials legitimized the next phase of clinical research.

Building on earlier success of using the sanction of scientists and physicians to promote the study, Hammon and NFIP officials solicited testimonials for use in further trial publicity. Whether advertising tobacco or beauty products, harnessing testimonials was a common tactic for marketers seeking to evoke a "personal approach," which often blurred the line between opinion and fact.[8] Tobacco firms, such as R. J. Reynolds, ran advertisements in medical journals featuring physicians who claimed that they found "'not one single case of throat irritation' from smoking Camel cigarettes."[9] Following a similar approach, Dorothy Ducas sought endorsements from Utah community and health leaders, which could be published to justify both the completed pilot study and the impending mass trial. In his testimonial, county medical society president Dr. Roy Hammond allayed concerns about health risks by explaining that "we have had no reports from the parents or the children of any difficulties or serious reactions in the nearly 6,000 inoculations that were given."[10] Similarly, state medical society president Dr. Vivian Parley White testified that his experience working with Hammon was so positive that "I am sure that any community in our State would be happy should you select it for a similar study."[11] The health commissioner, Dr. George A. Spendlove, recounted that "the experiment has increased our rapport with both medical profession and laity."[12] The Utah testimonials celebrated participation in the fight against polio and provided useful endorsements for later marketing purposes.[13] "With this background of experience," Hammon observed, "plus the photographic and tape-recorded records of the reactions of physicians, leading citizens, and parents, it will be much easier to sell this in areas to be selected for future tests."[14] Tangible evidence of solidarity and success were important tools for allaying the concerns of health professionals at future test sites.

From Pilot Study to Mass Field Trial

Hammon's plan to expand the GG study was given a boost by a series of laboratory findings in late 1951 and early 1952 that upheld the value of passive immunization. Yale University epidemiologist and virologist Dr. Dorothy Horstmann demonstrated that the poliovirus reached the nervous system through the bloodstream.[15] This discovery was independently verified by a Johns Hopkins University researcher, Dr. David Bodian, who confirmed the existence of a "transient middle phase" of polio infection during which antibodies in the bloodstream could neutralize the virus.[16] Bodian's research supported the use of GG, as the "paralytic disease could be prevented by the injection of much smaller amounts of Red Cross gamma globulin than those previously required."[17] In addition, University of Pittsburgh immunologist and pathologist Dr. Frank Dixon discovered that the "half-life of gamma globulin" in children "appeared to be about twenty-one days," which corroborated Stokes and Hammon's assumption that GG could last weeks.[18] Laboratory discoveries appeared to bolster the application of GG for polio and a need to continue with the field trial.

Despite optimism, the development of a competing immunization process appeared to challenge Hammon's commitment to GG. Dr. Jonas Salk, Hammon's chief rival at the University of Pittsburgh, was making inroads toward a prototype polio vaccine. Hinting at his progress in December 1951 at the Committee on Immunization, Salk stated, "I know that we tend to be enthusiastic about our own ideas and the things we do ourselves, but perhaps I may leave just a few figures with the group for consideration."[19] He presented data showing how monkeys receiving his new vaccine were protected from polio paralysis. Although not all committee members believed that his method would be effective in humans, Salk began a small series of human trials in the spring of 1952 with residents of the D. T. Watson Home for Crippled Children and the Polk State School.[20] Compared with a vaccine, GG appeared to be an imperfect solution because it was expensive, in short supply, and was needed for the control of other childhood diseases. Salk's competing discoveries threatened Hammon's approach, but did little to shake Hammon's belief in GG.

Identifying a community to host the next phase of the experiment was difficult. Michigan statistician Dr. Fay Hemphill and three epidemic observation teams grew frustrated with months of disappointment in locating a test site among 177 areas under surveillance.[21] As May turned to June, they learned that the residents of Houston and surrounding Harris County, Texas, were contending with one of the worst polio epidemics in the region's history.[22] The paralysis rate was 27 per 100,000 people, with dozens of new cases reported each week.[23] Although Hammon preferred to locate his study in a small town, he began to consider Harris County as time appeared to be running out.[24] Despite a large county population of over 800,000 people (640,000 in Houston), the absence of an alternative community combined with Harris County's expansive medical facilities and familiarity with GG during the 1948 epidemic made it an appealing proving ground.[25] On June 24, Hammon visited Houston "for a fast check to determine finally whether the test would be made."[26] After private consultations with the county medical society, where he shared the testimonials gathered from Utah County, Hammon became convinced that local support for the experiment would be strong. Concerned that a more suitable polio epidemic would not materialize, Hammon decided to remain in Houston and launch the study.

Hammon and NFIP officials applied the lessons they learned from Utah County to Harris County. They centralized operations and recruited supervisors to coordinate volunteers and clinics. NFIP southern regional director Robert Pierce was recruited to serve as chief operations officer, while NFIP chapter volunteer R. O. (Helen) Pearson was enlisted as director of women volunteers. Staff training was also improved; clinic personnel were given lectures and practiced at a "dry run" of the clinics in Pittsburgh. Hammon introduced preloaded disposable syringes to speed up injections, reduce the chance of contamination, and eliminate the logistical challenges related to sterilization.[27] The marketing program was also refined. NFIP officials developed a sophisticated publicity pack, complete with sample radio interviews, church endorsements, school addresses, and newspaper editorials.[28] To recruit the children most vulnerable to polio, Hammon amended his protocol to enroll those between the ages of one to six years, instead of two to eight as in Utah County. Based on this age range, he hoped that Harris County would provide thirty-five thousand child subjects.[29] Protocol changes and new equipment promised a smooth operation.

As with Utah, marketing played an important role in building support for the study. The approval of county doctors and health officials was used to sell the program to other groups. As women's clubs were an important source of family influence and volunteer labor, Helen Pearson organized "a meeting of the presidents of all women's clubs in the city."[30] Thousands of women volunteers were organized to promote the study and serve as clinic assistants. Embracing new broadcasting technology, Hammon agreed to a television interview, and according to one observer, he "seemed to enjoy it thoroughly." Meanwhile, NFIP personnel cultivated "very friendly and close relations with the working

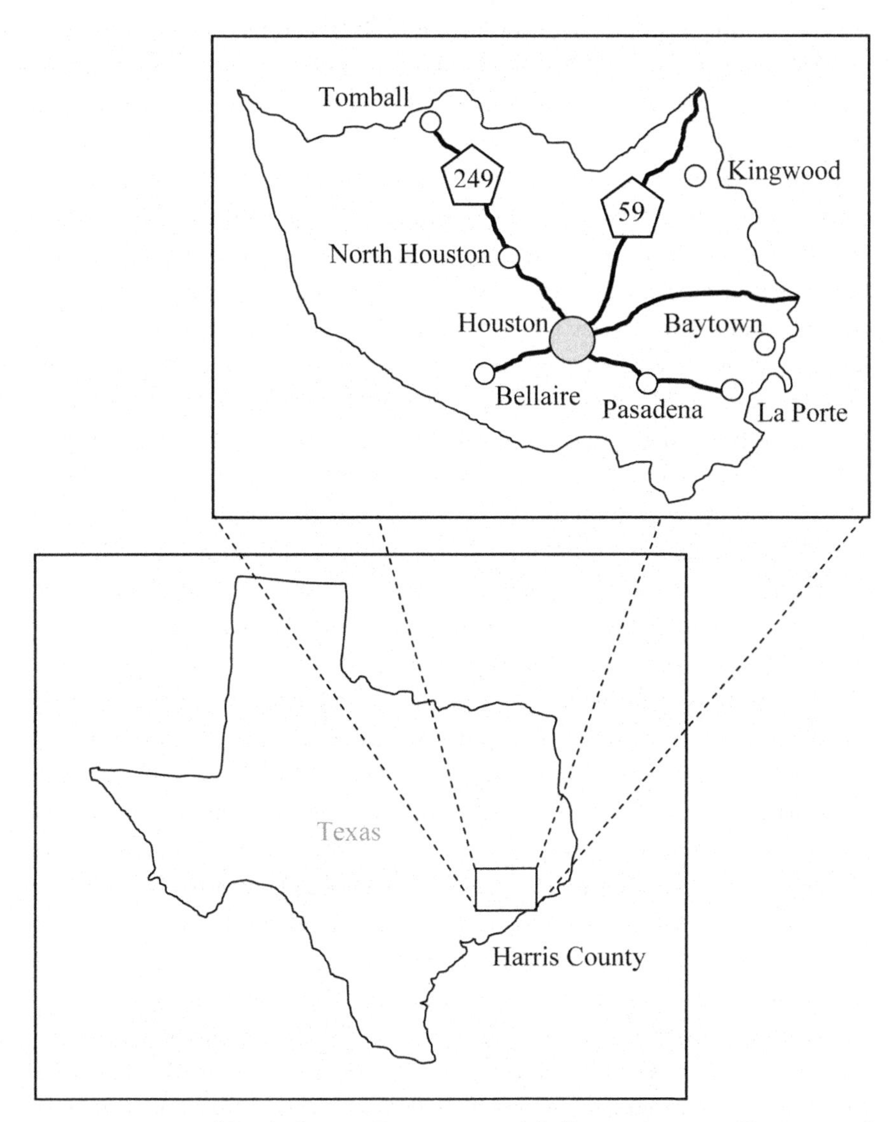

Figure 16 Map of Harris County, Texas, gamma globulin test site, 1952. Figure created by author.

press" and supplied journalists with "quite a bit of straight reporting" on the program. As NFIP regional director Frank Chappell noted, "There has been no problem of inducing the press to run stories—rather it's been a problem of keeping them supplied with enough data to cover sometimes as many as six stories in a single issue of one paper." While local publicity was encouraged, the NFIP took care to restrain national publicity. *Life* editors caved to pressure to limit coverage of the event, while NBC reporters agreed to abandon their planned special segment on the Houston clinics. Carefully managed publicity increased interest in the study, while shaping how it was portrayed in the media.[31]

Harris County doctors were encouraged to be among the first to volunteer their children to the study. Rosemary Burnett and her husband, Dr. Matthew Burnett, a local physician and member of the county medical society, enrolled their five-year-old daughter Patricia Ann as "the first child" to receive the injection.[32] Patricia Ann was also interviewed at the local radio station as part of the medical society endorsement. "Do you know why you're here, Patricia?" the radio interviewer inquired. "Yes," she answered, and Rosemary continued, "I think the effectiveness of gamma globulin is something we should all try to find out in this emergency."[33] Local radio stations were primed with such coverage by NFIP radio director John Becker, who was "flooding the air ever since he hit town."[34] The publicity surrounding the enrollment of local doctors' children was an effective means of shaping public perceptions of the study. Freighted on experiences from Utah County, Hammon and NFIP officials used doctors and the local media to build local support for the study.

As a result of the media campaign, misconceptions over trial leadership strained the relationship between Hammon and his sponsors. A growing awareness of NFIP involvement in the study led some community leaders to welcome Hammon as an NFIP employee. "Hammon has on a number of occasions been introduced as from the National Foundation," one NFIP representative observed, "and he didn't like it." Beyond embarrassing introductions, "Hammon was somewhat disturbed Sunday and Monday that the NFIP was getting too much credit." Since Hammon wanted his experiment to be perceived by the public as an independent research program, he contacted media outlets "and asked them to cut out NFIP from the stories." NFIP officials maneuvered to counter his efforts by "specifically asking reporters who call in for news each day to give us a mention."[35] Although Hammon and the NFIP jostled for acknowledgment in a seemingly optimistic moment in the fight against polio, their tactics nurtured resentment.

In the oppressive muggy heat of July 2, 1952, eight injection clinics were opened at public buildings throughout Houston to permit the enrollment of thousands of children. Drawing on the lesson of recruiting religious leaders, NFIP officials arranged for the president of the Greater Houston Council of Churches, the Reverend Thomas W. Sumners, to bless the opening of one clinic. Sumners intoned, "O Gracious God . . . Grant Thy blessing on their experiment . . . that the scourge of polio may be removed."[36] As anticipated by Hammon, the community response was tremendous, as anxious parents reacted favorably to the publicity program and eagerly lined up at the injection clinics.

"My first reaction was that I didn't want my child to be a guinea pig. But then I got to thinking," one Houston mother explained to reporters. "It can't hurt them, so we haven't lost a thing in coming."[37] The promise that participation was inherently safe made the 50 percent chance of a child receiving a potentially protective substance appear rational. As parents, some local doctors offered their close relatives to the study. Dr. Fred Laurentz brought his grandchild, David, to the clinics. Recalling this moment, David's mother Carleen Laurentz reflected that she and her husband "were willing to let our son take part in the field test because we trusted Dr. Laurentz's knowledge, expertise and judgment. We hoped the gamma globulin might have some benefit. I don't recall how David reacted to the inoculations. He probably cried a little, but he loved his grandfather, and would do anything to please him."[38] Besides encouraging the participation of family members, doctors were eager to offer assistance in conducting injections. "Some of the doctors are even protesting

Figure 17 Gamma globulin study injection clinic, Houston, Texas, 1952. Courtesy of the March of Dimes Foundation.

that there aren't enough shifts available in the clinics for even more than the 170 already signed up to work," one NFIP official noted.[39] The interest of Harris County residents buoyed the confidence of Hammon and suggested a favorable outcome.

Although parents and volunteers were keen to participate, the injection clinics tested their endurance. In some instances, lines exceeded what could be processed during opening hours. Many families waited hours, and three clinics became "so crowded with anxious parents and their children that officials were forced to send some of them home."[40] For parents considered fortunate enough to reach the injection station, their children often protested vigorously. "The doctors are using big, heavy needles about 2½ to 3 inches long, and the kids are really yelling," one NFIP official observed. As summer temperatures in Harris County peaked, congested lines and stuffy clinics became excruciating. "With the exceptions of two spots that are air-conditioned," one witness reported, "the heat is fierce. What with the waiting lines, intense heat, yelling kids, and milling parents, there was plenty of action." Under such conditions, clinic volunteers grew exhausted. "About half of the kids," remarked a clinic observer, "fight like mules and the doctors and nurses have to hold them down bodily. In Texas' July heat, this gets to be hard work after eight hours."[41] Despite the conditions, parents and clinic volunteers retained their commitment to the program.

The study exposed Harris County children to a range of adverse health reactions. The pain caused by large dose injections led some children to fall or lose consciousness. Of four serious cases, two children experienced "collapse, rapid pulse rate, shallow respiration, and pallor." Another child "became comatose and cyanotic [bluish coloration of the skin] and had a generalized convulsion while still on the injection table." As doctors since the 1920s had turned to epinephrine (adrenalin) when faced with cardiac arrest or anaphylaxis, an injection of this substance was swiftly administered.[42] When the child "walked away feeling fine," Hammon expressed no further concern. Allergic reactions also affected ten children, and in one case the child "began with coughing followed by edema of the face and later generalized urticaria [hives]."[43] Although Hammon noted similar reactions in Utah County in 1951, he considered them unusual and of minor risk.[44]

Despite trial administrators' success at building and maintaining community support, they faced many challenges adapting to a large city. "The entire preliminary five days was complicated by the fact that the entire operation was planned for a city of 100,000 and Houston has close to 800,000," observed

Frank Chappell. "It was necessary to set up eight clinics instead of five, as originally planned, and this complicated problems of supplies [and] volunteers. . . . Virtually nothing went exactly as planned in the dry run in Pittsburgh." As the scale of the experiment increased, confusion mounted; some clinic schedules and locations became mixed up and parents were turned away. Some intrepid journalists sought to assess parental uncertainty. One reporter with the *Houston Post* visited clinics and asked waiting parents, "You don't think this is any good, do you?" NFIP representatives monitored the reporter closely, assuming that he had "an assignment to look for trouble."[45] Even though competition between newspapers fueled controversy, such coverage did not undermine public confidence. It became evident to Hammon that conducting a medical experiment in a large city was a challenging enterprise.

Extending the experiment to Texas forced the field team to negotiate legalized segregation and the politics of Jim Crow.[46] Even before the test site was selected, NFIP personnel imagined how segregation might affect the study. "Segregation problems should be considered early in the picture," Pierce warned.[47] NFIP officials considered the enrollment of African Americans essential to upholding the pledge of fighting polio irrespective of race, but that clinics should not challenge legal norms.[48] Even though GG was derived from blood donated by people of many races, and data amassed from the trial would not be racially assigned, the field trial would need to recognize separation of families based on skin color. "It would appear that the best plan would be to set clinics up in both white and Negro schools," Pierce suggested.[49] NFIP director of "Negro Activities," Charles H. Bynum, served as an important adviser on the issue.[50] Bynum was formerly an educator in Texas and remained a champion of civil rights issues.[51] He was hired by the NFIP in 1944 to assist with fund-raising in black communities, but as his department expanded, he helped negotiate segregated practices and assisted Hammon and Pierce with "locating clinics" at local schools and community centers.[52] For Bynum, the inclusion of black Americans under segregated conditions was preferable to exclusion.

The difficulties associated with a segregated medical experiment tested the determination of many black families. Out of eight injection clinics, only one was designated for blacks at the Jack Yates High School.[53] This was justified by Hammon on the basis of Harris County's racial demographics, in which whites constituted a majority.[54] Meanwhile, a second injection clinic was opened to black and white families on a rotational basis at the Lyons Health Center.[55] White families were invited to attend the first days of the experiment, whereas black families were asked to "report" on alternate days near the end.[56] Black families faced later inoculations, fewer enrollment days, longer clinic lines, and restricted geographical clinic distribution. Confusion over clinic locations and dates, combined with a need to travel many miles to attend the clinics, placed a considerable burden on black parents who wanted their children to participate in the study.

Like many Harris County white parents, black parents not only volunteered their children, but also promoted the study and assisted at the clinics. Community leaders were amazed at the turnout and appealed for local volunteers to assist clinic staff.[57] "The swamped clinic at Jack Yates High school," the African American newspaper *Houston Informer* reported, "was calling for more volunteer workers before noon."[58] African American volunteers served as clinic clerks, line managers, and ushers, while black doctors administered serum injections alongside members of Hammon's team. This volunteerism helped transform the study into a community program, rather than an experiment designed in Pittsburgh and conducted by white doctors. Like the U.S. Public Health Service study of untreated syphilis in African American men in Tuskegee, Alabama, the involvement of black doctors and nurses increased the trust of families and the sense of contributing to scientific knowledge.[59]

Black Americans became increasingly important to the study, as some white parents began to seek alternatives. Although Hammon knew that GG could be readily purchased in Houston, Texas, he hoped its use could be restricted and deliberately avoided mentioning GG by name or implying its potential value against polio.[60] However, once Harris County parents and doctors learned that the blood fraction might be efficacious, private demand for the serum soared. "Since announcement of its use in the city's great polio study," one local newspaper reported, "Houston's pediatricians have been deluged with requests to administer gamma globulin."[61] The experiment's marketing program was so successful that some white parents began to purchase private injections of the blood fraction.[62] "Doctors were put under terrific pressure from their patients," John Becker observed. "On the other hand," he continued, "some doctors even called up some of their patients to come down to their offices and get the real thing, charging some fantastic fees. I only hope the March of Dimes gets 10 percent of what these doctors made."[63] In some instances, white families not only enrolled their children at Hammon's clinics, but then undermined the statistical value of the study by having the family doctor "give their children a 'sure' shot" of the blood fraction afterward.[64]

Hammon turned to newspapers to dissuade this seemingly selfish practice that threatened to sabotage the experiment. "The accuracy of the study," the *Houston Chronicle* pleaded with readers, "rests on the shoulders of Harris County parents, on whom the entire world is depending to help science find an answer to polio paralysis."[65] When such efforts failed, Hammon relocated clinics away from noncompliant areas. The Baytown Clinic was removed from the roster "because of the probability of 'boot-legging' of gamma globulin."[66] Hammon knew that he needed steady enrollment to achieve statistical significance, but white parents' support for the experiment was in decline. "Unless attendance picks up considerably," one NFIP staffer noted, "the required 35,000 kids cannot be reached in ten days."[67] Pharmaceutical suppliers openly promoted their GG stocks to parents and doctors. "I would say that the demand for gamma globulin from physicians has increased four [to] fivefold recently," one pharmaceutical representative admitted.[68] Although many physicians refused to administer private GG injections, wealthier and predominantly white parents exercised considerable agency.[69] For these parents, it was preferable to pay three dollars per cc of serum to ensure their child received GG rather than endure stifling clinics, long lines, and a 50 percent chance of possible protection.[70] For Hammon, however, data generated from a number of white children were ruined by the actions of audacious parents.

Though a few black families may have purchased GG injections for their children, evidence suggests that most did not. Economic factors, combined with rising acceptance of the study, increased black participation during a decline among whites.[71] One journalist observed that "despite an over-all slump in attendance at [white] clinics Friday . . . the Lyon Clinic had its biggest day with 404 Negro children receiving injections."[72] The final day of the Jack Yates High School clinic set "a record for the number of daily shots given in the field study" with 603 injections.[73] As a result, Hammon extended the duration of the study by two days and added two black-only injection clinics.[74] Although the number of clinics still disproportionately favored white families, the participation of black Americans helped Hammon accumulate data of the required quantity and quality.

The field test in Harris County was a mixed success. Although an unprecedented 33,137 children had been injected when the clinics closed on July 12, there remained a shortfall of two thousand.[75] Moreover, a scarcity of follow-up personnel to monitor and evaluate emergent paralytic polio cases hindered data collection efforts. Worse still, data quality was affected by the widespread undocumented use of GG. Pharmaceutical sales reports for July showed that over 20,000 cc of the blood fraction was sold to Harris County doctors.[76] As doctors usually injected only 1 or 2 cc of the serum to their pediatric patients, one may assume that the scale of private injections was at a minimum of ten thousand doses. While Hammon anticipated some of these challenges, he was not prepared for the scale.

After failing to control the GG experiment in Houston, Hammon and NFIP officials were desperate to locate a more manageable community. In mid-July, surveillance teams descended on Sioux City and surrounding Woodbury County, Iowa, when evidence emerged that "a very large number of [polio] cases would occur within the next few weeks."[77] Sioux City public health officers reported 228 polio cases for the year and eight new cases admitted in one day.[78] Saint Joseph's and Saint Vincent's Hospitals were inundated with patients, many of whom required "iron lung" respirators to survive.[79] As Sioux City physician Dr. Don Wagner remembered, "All we did was concentrate on polio patients."[80] A Saint Joseph's nurse, Arlene Moltsau, recalled how "people were afraid to go places and many events were cancelled."[81] Local newspapers cited alarming statistics and recounted heartrending accounts of recent polio deaths.[82] Iowa resident and polio survivor Joe King remembered that the "doctor came to our farm and told my folks it looked like polio and I needed to be admitted to the hospital."[83] The public health crisis affecting Iowa captured the attention of Hammon and NFIP officials.

Besides an intensifying polio epidemic, Sioux City boasted a manageable population of 84,000 and a measure of operational privacy, with its location two hundred miles west of the state capital, Des Moines.[84] The county was nearly homogeneous in terms of race and ethnicity. The 1950 census reported that 99 percent of residents were white. Across the Missouri River, the neighboring community of South Sioux City, Nebraska, and surrounding Dakota County constituted an additional ten thousand residents. Hammon considered both communities "stable," since their populations had not changed for over a decade.[85] For the field team, Sioux City and South Sioux City represented an ideal choice for a proving ground.

On July 14, Hammon moved supplies and personnel from Texas and set about "to establish the clinics only in or near these two cities."[86] Like Hammon, NFIP officials were pleased by the choice of Sioux City and South Sioux City, since the population size seemed manageable and statistical significance was within sight. "Having just returned from a day in Iowa," John Becker explained, "I think the fact that the next field trial is going to be held there is a lucky one. The 'Community Leaders'—to coin a phrase—that I talked to, namely doctors, lawyers, etc., either had no idea of what we were using or thought it was unavailable for public use."[87] As observed in Utah County, the Iowa and Nebraska test site promised cooperative residents with limited knowledge of GG.

To avoid repeating mistakes, Hammon pressured local health professionals to refuse all private GG injections. During his presentation to the Woodbury County Medical Society, Hammon requested that doctors go "on record in favor of its members' giving no gamma globulin . . . except in our organized clinics." Hammon also received assurances from local pharmacists "to sell no gamma globulin" for the duration of the study. Like Utah County, the cohesive nature of the small-town community enabled Hammon a measure of compliance.[88] By anticipating citizen agency and requesting conformity with the protocol, Hammon averted the illicit allocation of GG.

On July 17, trial administrators unleashed their marketing campaign in Sioux City and South Sioux City. A press conference was organized during which Hammon described to reporters the nature of the experiment. The *Sioux City Journal* ran a front-page feature the following day. "If citizens respond in sufficiently large numbers," the article quoted Hammon, "we may be able to get to the bottom of this question once and for all. Otherwise, small scale tests will have to be continued indefinitely until a scientifically valid conclusion can be reached." Citizens were asked to participate out of duty and to help scientists reach an answer. To instill confidence, Hammon assured parents that he had observed "no serious reaction" among children at other test sites, and he convened special meetings to "answer questions concerning the project."[89] Editorials opined that the plan was "highly scientific" and safe thanks to the approval of local medical societies and health departments.[90] Large published photographs of arriving nurses shaking hands with Hammon or contract clinic staff posed in front of commercial aircraft added to the exciting atmosphere.[91] After four days of advance publicity, the marketing program had generated incredible enthusiasm.

As witnessed at previous test sites, Sioux City women helped promote the study and assist at the injection clinics. Timed with the release of promotional materials, Helen Pearson set about mobilizing local women.[92] She organized meetings to promote "the duties of the volunteers," and she "showed motion pictures of the Provo trial and colored slides of the Houston study."[93] Through this public education effort, many Sioux City and South Sioux City women enlisted as clinic assistants, supply clerks, and community promoters.[94] "The response was as good [as] or better" than Houston, Hammon reflected. One industrious Sioux City woman reportedly arranged for four telephone networks, or "party-line hookups," to be interconnected. This linkage enabled

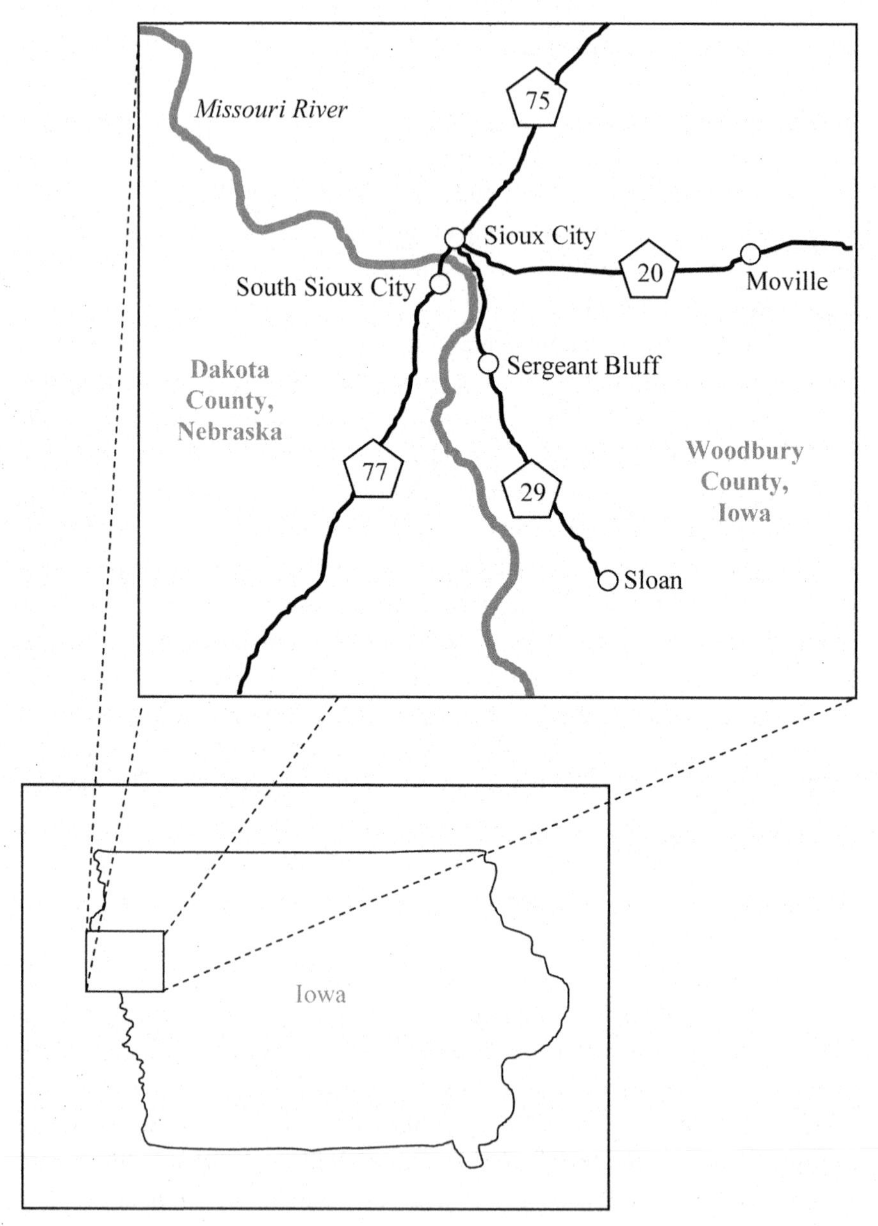

Figure 18 Map of Iowa and Nebraska gamma globulin test site, 1952. Figure created by author.

Hammon and his associates to reach over three thousand rural families simultaneously and explain to them the nature of the test and "where and how the listeners could participate."[95] The mobilization of women opened up new promotional opportunities and reduced logistical demands.

When the five injection clinics opened on July 21, thousands of parents clamored to volunteer their children. Since Hammon required over fifteen thousand subjects to achieve statistical significance, he amended the protocol to enroll those from age one to eleven years, instead of two to eight as in Utah County and one to six in Harris County.[96] This expansive range facilitated a significant public reaction. "The unprecedented response to mass inoculation tests at polio study clinics was expected to continue at Sioux City," one newspaper touted, "with hundreds more children receiving shots."[97] Like many parents, doctors clamored to be part of a history-making moment. One pediatrician took time away from his practice to help administer injections, while the son of a prominent Sioux City physician was called home from vacation to assist.[98] Hammon was pleased by the response, which he reasoned "surpassed the opening day's attendance in previous tests in Utah and Texas."[99] The enthusiasm of local families, combined with an experienced field team and an efficient injection method, permitted swift enrollment. By the end of the first day, clinic personnel had inoculated "a record-breaking 3,544" children, while an evening clinic "boosted the total to 4,317."[100] To meet the demand for clinic volunteers, the 79th U.S. Air Force Squadron sent troops to help at the clinics.[101] Military personnel, combined with citizen volunteers, doctors, and nurses, worked together at the clinics to fight the epidemic and generate a sense of hope.

The publicity program did not convince all Woodbury and Dakota County citizens to participate in the study. Evidence suggests that some families remained concerned by the risk of crowd contagion and refused to enroll their children in the study. One former child subject remembered, "My friend did not participate. Her parents thought she might contract polio in a public place where so many people would gather. She was pretty much home bound that summer."[102] Such parents saw the paradox of advising crowd avoidance while congregating children at injection clinics.

Despite the personal nature of injecting a serum into the gluteus maximus, little effort was taken by Hammon and his field team to shield children from public view in consideration of their privacy. As some Iowa children were older than those in Utah and Texas, they were more self-conscious of their bodies. "This was a very undignified experience for a girl looking forward to being eleven," one participant remembered. "There were many tables and each

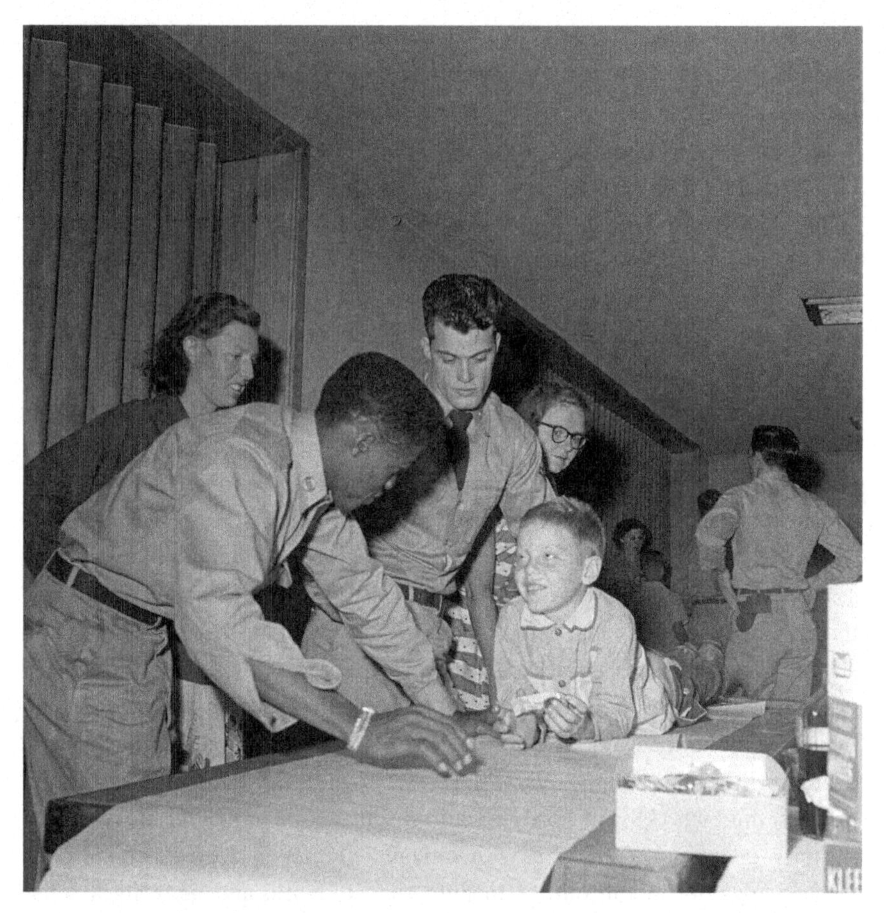

Figure 19 GI from the 79th Air Force Squadron gives a boy a lollipop after his injection, Sioux City, Iowa. Courtesy of the March of Dimes Foundation.

had a line of parents with their children and the children were not happy. The process involved dropping your undies, leaning over the table and having the longest needle I had ever seen plunged into my bottom."[103] Sharing this sentiment, another former participant remembered the impersonal clinic environment. "Lots of tables and lots of people with white on," he explained. "You lay down on the table face first. . . . They pulled your pants down over your butt and gave you the shot. No separation between boys or girls or ages. Line up, lay down, and take it."[104] The neglect of privacy, combined with the impersonal setting, contributed to children's psychological and physical discomfort.

As observed at other test sites, some children experienced adverse health reactions following injection. Several cases of fainting were noted "in boys aged nine to eleven." Although most of the older children received the largest 11 cc dose, based on their weight relative to infants, Hammon dismissed the issue by reasoning it was "of psychogenic origin." A few children also suffered fever, which he did not consider worrisome.[105] Although adverse reactions were a persisting theme during the study, Hammon was reluctant to acknowledge how they might complicate the future clinical use of GG for polio.

The Iowa phase of the study was a triumph for Hammon and his collaborators. When serum supplies were exhausted by noon on July 27, the clinics closed ahead of schedule. During the course of the study, local doctors refused to administer private doses of GG and local sales of the blood fraction were minimal. Many parents were pleased to help in the generation of scientific knowledge and believed that their children might be protected from the ravages of polio.[106] Hammon and his team had successfully inoculated 15,595 children in fewer than six days: a remarkable feat considering the predominantly rural context.[107] Most importantly, statistical significance was finally achieved with a total enrolment of 54,772.[108] Evidence provided by Iowa children completed the data set and allowed Hammon to assess the efficacy of GG in the prevention of paralytic polio.

Selling Scientific Outcomes to the American Public

Promotion of the GG study did not end when the Sioux City injection clinics closed. Well before the follow-up teams returned from Iowa, Hammon and NFIP officials began preparing for the next phase: marketing the outcome. Justifications were necessary for many reasons. Though pioneering in its application of controlled trial methodology, the experiment was inadvertently affected by the hand of human agency. It was impossible to know how much data was tainted by clandestine GG injections, multiple clinic enrollments, documentation errors, inconsistent injection practices, underreporting of paralysis cases, and variation of epidemic severity. Before tabulating the results from Utah, Texas, Iowa, and Nebraska, Hammon and his collaborators, Dr. Joseph Stokes Jr. and Dr. Lewis Coriell, needed to decide whether the data set was valid. Although they appreciated that several factors deleteriously affected the study, sources suggest they considered their data sufficient to assess the utility of GG for polio. After comparing the placebo and GG groups, Hammon and his associates concluded that the blood fraction offered three to five weeks of immunity against paralytic polio beginning the first week after injection. Of the 54,772 children included, 104 had succumbed to paralytic polio.[109] Within this group, thirty-one were recipients of GG while seventy-three received placebo. The scientists reasoned that the difference between the groups showed that GG prevented 1.59 cases of polio paralysis per 1,000 injections.[110] They were pleased by the results, which represented the culmination of almost three years of persistence and lobbying. It was expected that Hammon would publish his scientific outcome in a leading journal to gain recognition and disseminate knowledge. Many doctors, public health officers, and science writers were eager to learn of Hammon's results. Critics of the study, such as Minnesota's Dr. Gaylord Anderson and New York's Dr. Herman Hilleboe and Dr. Robert Korns, also needed to be convinced that the experiment had not provoked polio or exacerbated the epidemic. Finally, it was important to the NFIP that March of Dimes donors were convinced that sponsorship of Hammon was a responsible use of funds.

Hammon, Stokes, and Coriell published their findings in the *Journal of the American Medical Association* (*JAMA*). As a leading periodical and official organ of the American Medical Association, *JAMA* assured Hammon and his collaborators a wide readership within the United States and abroad. *JAMA* articles, with the scrutiny of an editorial board and oversight of referees, were also considered rigorous and of high quality. Through a series of four installments, Hammon and his colleagues justified their experiment and sold its results. They explained that the study was necessary because "epidemiological studies had led to no hope of [polio] control in the near future" and "no active immunizing agent was available."[111] Since GG was a licensed substance used to control other diseases, it was framed as inherently harmless. Although some public health leaders and members of the Committee on Immunization criticized Hammon's plan, the *JAMA* articles suggested that his protocol was uncontested and a logical outgrowth of scientific inquiry.

The *JAMA* articles framed the GG study as an independent venture undertaken by dedicated scientists. Although NFIP officials, consultants, and chapter volunteers played important roles in planning, promoting, and conducting the field trials, their contributions were not discussed, but merely relegated to a footnote, as only purveyors of funding.[112] The volunteer contributions of Utah, Texas, Iowa, and Nebraska parents and their children were similarly ignored. As was common in medical literature, Hammon sought to emphasize the scientific elements of his study to increase its merit.

The authors accentuated how their study was lawful and respectful of parental rights. The involvement of local doctors was linked to their authority and state licensing laws. The parental consent form was exhibited to showcase the efforts to enlighten parents about the study and to obtain their written consent. The authors suggested that the form was provided for educational purposes so "that all parents could be fully informed and understand."[113] Although the form did not meet the 1947 Nuremberg Code standards for informed consent, it was situated by Hammon, Stokes, and Coriell as a legal triumph and an achievement in clinical trial transparency.[114] Their openness about the legal dimensions of the study helped deflect attention away from ethical quandaries.

The *JAMA* article attempted to show that the GG study was harmless. The authors described efforts "to avoid crowding" at the clinics and "to keep family groups isolated" during the epidemic. The large-dose injections were presented as "practical" and the maximum amount "that could be given without producing too much discomfort or disability." The possibility of serum hepatitis was recast as a latent discovery that had been appropriately addressed. Evidence of adverse health reactions, including collapse and allergy, were framed as unavoidable facets of injecting thousands of children. The authors skirted polio provocation by explaining how the study provided an opportunity to measure the risk. "It was encouraging to note," they concluded, that "there was no suggestion here that the inoculation of gelatin had induced any paralysis."[115] Through the article, the authors fashioned the experiment as a cautious undertaking that considered all health consequences.

Hammon, Stokes, and Coriell also attempted to demonstrate that their study was scientific and statistically rigorous.[116] Since accuracy was a hallmark in the generation of scientific knowledge, the authors emphasized the application of placebo controls and the statistical underpinnings. The equation and the significance chart were published, implying that both method and data were sound. The prominent inclusion of graphs and comprehensive tables further accentuated the aura of accuracy.[117] Even though the clinical trial was plagued by unknown variables and human agency, it was framed by the authors as impartial and valuable.[118]

In addition to accuracy, the authors claimed that the study was merited. Since GG was among the first immunizing agents to be evaluated in a large controlled trial, Hammon, Stokes, and Coriell were proud of their accomplishment. "Had typhoid and rabies vaccines been originally tested in this manner," they boasted, "years of uncertainty might have been saved and we might not be in out present, unenviable quandary regarding the efficacy of these immunizing procedures after a great many years of routine application."[119] Although the authors admitted that GG was no "panacea," they framed it as a pioneering step in conquering polio. Since the experiment could not be easily duplicated, readers had to trust that the information conveyed by the authors was complete and accurate.

Not all researchers were convinced by Hammon's results. NFIP scientific adviser Dr. Thomas Rivers was privately concerned by data quality and later explained in an interview with Saul Benison that although "close to six times as many children were inoculated with gamma globulin and placebos in Houston as in Provo, the tests did not have the validity they should have had, simply because the doctors and the public in Houston did not play the game according to the rules." For Rivers, the Houston data set was unreliable, and this fact ruined the study. "It bollixed up the results of the trial good and proper," he asserted. "I hate to say this about doctors in Houston, but they shouldn't have done this. They thought they were doing good, but they weren't."[120] Although he questioned the accuracy of these data, he did not openly criticize the study out of concern for Hammon, the NFIP, and Houston doctors.[121] "I didn't have to wait for the results of that test to know that they wouldn't come up with anything," Rivers later reflected. "I'll admit that to this day I have never bothered to read the final report of that test. It's written up somewhere. You read it and if it proved anything I'll eat it for lunch tomorrow."[122]

With no open acknowledgment of the serious problems that had plagued Hammon's GG study, the media rushed to celebrate the blood fraction and the dawn of temporary polio protection. *Popular Science* revealed that Hammon "proved beyond doubt the power of gamma globulin."[123] Under the headline "Gamma globulin to the rescue," *Life* magazine proclaimed that GG "gives good . . . protection against paralytic polio."[124] Local newspapers, such as the *El Paso Herald-Post*, informed readers that the blood fraction "cuts [the] likelihood of crippling; protection lasts five weeks."[125] As the public learned of this new medical breakthrough, Hammon was catapulted into the world of celebrity. As medical writer Richard Carter observed "the press began fitting Hammon for halos."[126] Medical societies and public health departments invited him to speak about polio prevention, and reporters requested interviews.[127] Recounting the exploits of medical heroes, the publishers of *True Adventures of Doctors* popularized Hammon's victory in an inspiring account for children.[128] Hammon emerged from the medical experiment to become America's leading polio warrior.

NFIP officials sought to share in the glory. A conference was organized in November 1952 at the New York Waldorf Astoria Hotel where Basil O'Connor announced the "successful results" of Hammon's study. "Man has scientifically prevented paralytic polio in human beings for the first time in history," he asserted dramatically. O'Connor explained that GG was not the solution to polio, but an important step from basic research to "solid ground."[129] With the assistance of NFIP publicists, *Popular Science* made it clear that the study was "backed by the March of Dimes," whose support of research was "Closing in on Polio."[130] Although NFIP officials trumpeted their financial contributions to the study, they did not disclose their incredible operational and publicity support out of regard for Hammon's need to appear independent.

The NFIP chose to commemorate the study with a short film, titled *Operation Marbles and Lollipops,* which was to be shown on television and cinemas. The NFIP had considerable experience using film for public education and March of Dimes promotion. Like earlier NFIP films, *Operation Marbles* fused a collective hope for conquest over polio with an emotional appeal for donations. In the style of many 1950s civil defense films, *Operation Marbles* featured an upbeat narrative spliced with compelling footage of Hammon's study and the reassuring commentary of newsreel narrator Ed Herlihy.[131] Designed to link Cold War notions of community and patriotism, the film was a masterpiece in public relations drama that portrayed the GG study as an important social, economic, and medical achievement.[132]

Operation Marbles upheld public assumptions of medical research as rational, safe, and valuable. Opening with footage of NFIP assistant director of research Dr. Henry Kumm, the film dramatized the process of randomizing the gelatin and GG ampules by drawing black or white marbles from a box. "This is no children's game," the narrator stated, "This is serious business."[133] The process of designing and instituting a controlled trial was shown to be essential to science. This was "as scientific and impartial as it could have been within the walls of a scientific laboratory," the film boasted. The protocol was presented as a rational guide that was meticulously upheld. "Strictest clinical procedure was followed," the narrator promised, as the film showed clinic volunteers at their stations processing orderly lines of parents and children. Through an idealized rendition, the GG study appeared to be moving toward inevitable triumph.

NFIP producers went to great lengths to portray participating families as informed actors and not exploited subjects. Criticism of celebratory medical films such as *Mechanics of the Brain* (1933) or *Yellow Jack* (1938) had inspired 1950s producers to incorporate the theme of free will when discussing human studies.[134] "The parents of six thousand children were glad to cooperate with the visiting scientists," the narrator assured, as the film panned to a Utah County injection clinic, "even though they knew that under the conditions of a controlled scientific test their children stood only a fifty-fifty chance of receiving the blood fraction." For producers, it was important that the film show participation based on consent without coercion. Parents were shown smiling, guiding their children through the clinics, or helping to restrain their charges on injection tables. "All the kids got for their pains was a lollipop," the narrator stated, "or at least that's how it seemed to them." The ease of parental participation and their positive sentiment was shrewdly constructed in the film to counter criticism that a vulnerable population may have been "subjected to science" in a moment of crisis.[135] In portraying Houston and Sioux City, the film became increasingly fixated on freedom of choice. Scenes depicting families seated at registration desks, speaking with nurses, and consulting with doctors implied that informed consent and transparency had been upheld.[136] Commentary and imagery contrived an unambiguous argument in which parents not only accepted a controlled study, but also believed it was rational for their healthy children. Although youngsters had not volunteered themselves, the film steered clear of such issues by implying that parents knew what was best. "Even if some of the kids didn't think much of it," the narrator explained, "their parents understood though; that if even a few children were protected from paralysis by the gamma globulin, this controlled test was worthwhile. It did prove worthwhile."[137] Parents prevailed in their roles as legal arbiters on behalf of their offspring.

Operation Marbles' producers also used the film to sanitize the encounter between scientists and research subjects.[138] Although the experiment was characterized by long lines, crying children, and large needles, a less provocative version was presented. In one scene, a beaming mother was posed with her photogenic daughter on an injection table; meanwhile, in the background, the doctor commenced the injection. Once the procedure was completed, a lollipop was swiftly passed to the child. She and her mother looked up and smiled to the camera as the narrator explained: "This was science in action. But these were not mere numbers on a chart. These were America's children." Through this sentimentalized encounter, researchers were shown to be compassionate and concerned with the welfare of children.[139]

NFIP producers also downplayed the problems posed by racial segregation. Although black Americans in Houston were required to attend separate clinics, the film avoided such distinction so as not to rouse criticism at a time of growing civil rights activism.[140] Motion picture supervisor Joseph Cramer explained his frustration to Hammon: "The segregation problem, etc., had a great influence and placed severe restrictions on the editing of much of the film for TV release."[141] Akin to the race relations approach championed by Charles H. Bynum, film producers attempted to characterize blacks as equal partners with whites.[142] In one segment, a black child was shown wearing a straw cowboy hat and leaning on an injection table. Moments later, his teary demeanor was altered when presented with a lollipop. Themes of interracial cooperation were accentuated: both white and black doctors and nurses were shown working together at clinics registering and injecting children. The racially inclusive nature of the clinics was framed as not only a victory for science, but also a victory for race relations.[143] By minimizing depictions of segregation and the realities of racial prejudice, producers attempted to reinforce the NFIP commitment to all Americans, irrespective of race.

Operation Marbles was an important NFIP publicity vehicle used to share a breakthrough that "brought the fight against infantile paralysis much closer to its goal." As laboratory workers from the University of Pittsburgh were shown onscreen, the narrator explained that Hammon had "planned the tests under a grant from the National Foundation for Infantile Paralysis with March of Dimes funds." To reinforce notions of progress, producers contrasted the ideal of polio prevention with the indignity of polio treatment. As dramatic scenes presented "luckless youngsters" undergoing physiotherapy or treatment in iron lungs, the support of science and its quest for prevention by "seeking and testing" appeared logical and important. Following characteristic March of Dimes tactics designed to evoke emotion, the final scene focused on a photogenic child peering out from an iron lung; as she smiled longingly into the camera, the narrator asserted that through the GG study "marbles and lollipops can become the normal playthings of children the world around." The NFIP film complemented Hammon's *JAMA* articles and helped sell the experiment as a scientific victory. For the first time in American history, a medical intervention was shown to prevent paralytic polio in a clinical trial. Although Americans welcomed the promise of GG, it was uncertain whether it would become a valuable weapon in the armory of public health.

Chapter 6

The National Experiment

In August 1953, mere days before a summer camp near Livingston Manor, New York, was set to close for the summer, polio struck a sixteen-year-old resident. Worried parents collected their children and set about locating the much-touted liquid gold of polio prevention: gamma globulin (GG). When parents learned that GG supplies in New York State were "practically exhausted," they lay siege to the city health department. Led by Hyman Zarett, a real estate executive from Queens, the crowd of parents demanded answers as to why there was "such a shortage" of GG and requested the immediate release of supplementary supplies. This was far from an unprecedented event, and parents in neighboring counties stormed public health offices for the same reason.[1] One former New York resident remembered the crisis: "Everyone lined up to get very painful injections of gamma globulin. . . . They begged for double doses."[2] A swift appeal to federal authorities enabled the release of additional doses. By evening, New York parents lined up for doses of the precious blood fraction. Although Zarett and his fellow parents exercised a form of civic health activism, they also participated in America's first national polio immunization program. How did the national program come into being and how was it conducted? What were the results and how did they affect future plans? This chapter explores the program's successes and failures, as well as how the outcome affected the testing of the first polio vaccine.

The Hopes and Hardships of the National Program

The encouraging results of Hammon's GG study inspired preparations for a national public health program to control polio.[3] Newly inaugurated

Republican president Dwight D. Eisenhower was enthusiastic about the venture and assigned the matter to the government's scientific policy arm, the National Research Council (NRC), and to its logistics arm, the Office of Defense Mobilization (ODM). The ODM was a powerful Cold War organization established in 1950 to coordinate logistical activities for the Korean War and ongoing defense against the Soviet Union.[4] Since GG, as a blood fraction, was considered a valuable wartime asset, its processing, stockpiling, and distribution fell under ODM control.[5] The NRC and ODM enlisted the help of other agencies.[6] The American National Red Cross (ARC), the charitable, nongovernmental organization that provided GG to Hammon's study, was recruited to supply doses of the blood fraction.[7] Meanwhile, the NRC established a special advisory panel to create an allocation policy and usage criterion.[8] The panel comprised representatives from stakeholder organizations, including Dr. W. H. Aufranc from the ODM, Basil O'Connor and Dr. Harry Weaver from the NFIP, Surgeon General Dr. Leonard Scheele and Dr. Paul Wehrle from the United States Public Health Service (USPHS), Dr. Alexander Langmuir as director of the Communicable Disease Center (CDC), Dr. T. P. Murdock from the American Medical Association (AMA), and Dr. James Shannon from the National Institutes of Health (NIH).[9] Hammon was also invited to serve on the panel and assist its members in designing America's first polio immunization program.

The NRC panel favored an expedient GG allocation method and usage criterion. Serum supplies would be dispensed from a national stockpile managed by the ODM and distributed through the USPHS to each state health department.[10] Panel members reasoned that using existing public health infrastructure was an efficient means of moving serums during an emergency. The ARC agreed to fractionate blood from military and civilian sources at a cost of $3.5 million, while the NFIP promised to purchase commercial stocks of GG and cover shipping.[11] When polio epidemics struck, the ODM would deliver GG to staging areas in stricken communities. To improve the response time, initial allocations were extended to state health departments based on volumes of 1.5 cc of GG times the number of confirmed polio cases in the state since 1948. Out of an anticipated seven million cc supply of GG in 1953, over 60 percent would be allocated for polio control, while the remainder was reserved for measles and hepatitis.[12]

NFIP officials faced challenges mobilizing industry to supplement ARC supplies. First, GG was expensive, and unlike the ARC, commercial firms produced it only at a profit. However, the NFIP was running a deficit in 1953; the prospect of committing $5.5 million to purchasing and shipping GG, as well as a further $19 million in 1954, posed a hardship.[13] Second, as there was little demand for large quantities of GG before 1953, most pharmaceutical firms had not invested in blood fractionation.[14] New equipment, economic incentives, and technical knowledge would be needed to boost output. Third, since commercial firms fractionated GG from much smaller donor pools than those of the ARC, some researchers feared that polio antibody levels might be too low to convey protection.[15] Despite the expense and uncertainty, NFIP officials considered the national program an institutional priority and contracted leading firms, including Armour, Courtland, Cutter, Hyland, Lederle, Pitman-Moore, Sharpe & Dohme, and Squibb.[16]

Concerned by the impending GG shortage, the NRC panel selected a conservative usage criterion.[17] Whereas Hammon evaluated GG for mass prophylaxis and advised continuing this method for the national program, most panel members disagreed. Since it was difficult to predict the location and severity of a polio epidemic, and since GG needed to be injected one week before infection, most panel members reasoned that mass prophylaxis was logistically burdensome, expensive, and wasteful.[18] Instead, they advised using GG in a sparing and targeted manner by offering it only to household contacts of polio cases.[19] Although they acknowledged "multiple cases" within a family group was "not a common happening," they believed that the household contacts method was a "rational basis" to distribute a precious substance.[20] The method promised to use less serum than mass prophylaxis, and it held "psychological appeal" because its parallels with immunizations for smallpox, typhoid, and diphtheria made it easy for families to understand.[21]

Despite the public relations and logistical benefits of the household contacts method, NRC panel members did not know whether it would be effective. Since GG was not evaluated by Hammon for household contacts, ODM officials assigned the CDC division of the Public Health Service to evaluate the efficacy of the national GG program.[22] A special committee was convened by CDC director Langmuir, who was under no illusions about the difficulties involved with assessing GG on a national level with an untested usage criterion.[23] Since a controlled clinical trial was not possible under the circumstances, Langmuir and his epidemiologists knew their assessment would not be "a carefully planned research study," but rather "an attempt to collect as much information as possible."[24] The national GG program would inadvertently be another experiment.

The NRC's decision to use GG for household contacts worried some polio researchers. "If the gamma globulin field trials had stopped after Sioux City, everything would have been all right," NFIP scientific advisor Dr. Thomas Rivers reflected during an interview. "But damn it, in 1953 the National Foundation, in collaboration with the National Research Council, decided to put on another trial to test gamma globulin as a public health measure by giving it to family contacts."[25] In disgust, Rivers recalled how he attempted to stop the program. "I then made it my business to try to talk Mr. O'Connor out of giving the support of the National Foundation to this new test. I told him as bluntly as I could that I thought this new test would in all probability be a waste of time, money, and effort."[26] Basil O'Connor was also worried that the household contacts method would fail to protect children and fuel a public backlash against March of Dimes investment in medical research. In April 1953, he publicly condemned the NRC panel's usage criterion and insisted that the only responsible use of GG for polio was in the mass prophylaxis method assessed by Hammon.[27] "We note with some concern," the *Washington Post* quoted a NFIP press release, "that in accordance with this plan, the greater part of the Nation's stockpile of this scarce material may be used in a manner for which direct proof of efficacy is lacking."[28] When O'Connor's efforts failed to change the NRC panel's decision, he demanded that the federal government take responsibility for the program and incur its costs. "We respectfully request," O'Connor wrote, "that the United States government purchase the supply of commercial gamma globulin."[29] In spite of O'Connor's petitioning and criticism, the Eisenhower administration refused to inherit the program. Frustrated by their lack of influence on public health policy, O'Connor and his personnel continued their attack. A *New York Herald Tribune* article described the NFIP as being stridently "opposed" to the household contacts method, since "there was not experimental evidence which indicated how effective the substance would be."[30] Following O'Connor's lead, NFIP medical director Dr. Hart Van Riper opined that the national program was a farce and that parents could achieve better protection against polio by keeping children rested, clean, and away from crowds.[31] NFIP officials struggled to distance themselves from a usage criterion they suspected would yield little protective benefit.

Although NFIP officials were concerned by the design of the national program, they appreciated it offered important benefits. Federal control over GG would reduce the extent of illicit provisioning and improve access to a scarce material.[32] "We don't want GG bought on a black market now and stored in someone's private icebox to be injected next summer," Van Riper cautioned.[33] One doctor in Westchester County, New York, tried to anticipate demand by placing an order for 10,000 cc of GG and protested when he received only 1,000 cc. NFIP officials learned of this incident and "discouraged" the doctor from attempting further purchases.[34] Moreover, federal control would help shield the NFIP from public demands. Since NFIP chapters used local March of Dimes donations to cover polio hospitalization costs, officials feared that some contributors would demand privileged access to GG. "I've heard of pressure," one representative reported, "being applied on our chapter people and on physicians to get gamma globulin for selected children in the spring."[35] To prevent such pressure, NFIP officials instructed chapter personnel not to "under any circumstances purchase gamma globulin."[36] It was hoped that federal control over the stockpile would offer a secure system to manage demand.

The notion that federal agencies, rather than state public health departments or family physicians, would regulate access to GG and coordinate a national stockpile was a delicate concept at a time of McCarthy hearings and communist smear campaigns.[37] To make the national program palatable to patriotic Americans, it followed orthodox medical and public health practices. Doctors were required to request GG supplies for their patients through health departments.[38] They could also levy fees for administering the injections. NFIP officials hoped that each family would incur this cost without seeking assistance; however, when a family could not afford the fee, the NFIP advised its chapters "to underwrite such a cost."[39] Moreover, the autonomy of philanthropic and private enterprise remained intact. The government would not purchase the GG or nationalize production, but rather receive it as a donation from the NFIP and ARC. The NFIP would cover all shipping costs, thereby eliminating any government subsidy.[40] Finally, the national program would be presented as a military venture. At the height of the Korean War, it was no accident that slogans such as "Operation Lollipop" or "Operation Ouch" were used to promote the program.[41] Although the national program was a federally managed initiative, its private supply and cost management thwarted accusations of "socialized medicine."[42]

While many organizations contributed to the national program, they were unable to design a unified publicity campaign. In January 1953, animosity erupted when the ODM organized a meeting with the NFIP and ARC to discuss joint publicity. One ARC official criticized the NFIP for promoting Dr. Jonas Salk's progress with his vaccine and requested it be played down because it "hurt the blood program."[43] They believed that blood donations would "lag" if the public thought that GG was no longer needed to fight polio. NFIP officials countered that they never suggested a polio vaccine was imminent and that such concerns only proved the ARC was selling its program "on the basis of a need for a lot of blood for gamma globulin." Affronted, ARC officials refused to cooperate further. In the words of NFIP public relations director Dorothy Ducas, "I am sure I did not make myself popular, but, at least, they knew where

Figure 20 Michigan Air National Guard unloading gamma globulin shipment, 1953. Courtesy of the March of Dimes Foundation.

we stood."[44] As a result, each agency pursued its own initiative. NFIP officials grew worried that separate publicity campaigns would create "confusion" and misrepresent their organization. "I understand the ODM is now hot and bothered about getting out a motion picture film on blood recruitment," one official noted. "I'd like to be included in the pictures and if we're not, I'd like to know about it so that we can take some measures of our own."[45] The failure of contributing agencies to cooperate increased tensions and set the stage for conflict.

The relationship between the NFIP and ARC slowly deteriorated. Fund-raising for polio was long dominated by the NFIP and the sudden encroachment of the ARC stoked competition and resentment. The ARC launched an aggressive advertising campaign in 1953, linking its role in keeping soldiers alive in Korea with its domestic role in "combating paralysis in child polio victims."[46] ARC officials approached NFIP chapter volunteers, requesting support for blood donations and fund-raising. The rivalry confused donors and infuriated the NFIP. "I am not going to any great length," one NFIP representative noted, "to impress upon you the quandary in which . . . Chapter folks are finding themselves."[47] NFIP officials were aware of the implications of competition with the ARC and forbade volunteers from extending assistance.[48] In response, they launched a counter-publicity campaign that marginalized ARC contributions and emphasized the NFIP.[49]

The ARC exploited its political connections to claim the national program as its own. The mayor of New York City was approached by the ARC to offer testimony that "the Red Cross polio project" was important "for every parent to support."[50] The month of March was designated "Red Cross Month" and citizens were urged "to support without reservation."[51] Eisenhower, in the capacity as honorary ARC chairman, added further authority; he asserted that support for the campaign would allow the charity "to protect children against the paralyzing effects of polio."[52] Akin to Eleanor Roosevelt's support for the NFIP, Mamie Eisenhower reminded the nation's mothers that ARC donations would permit "a new weapon" to "prevent the crippling and deformity" of polio.[53] The hope that a blood fraction could control paralytic polio had enabled the ARC to access the coveted NFIP donor base.

Public Relations and the National Program

When plans for the national GG program were unveiled to the public in the spring of 1953, NFIP officials reminded Americans of their prominent role in its implementation. The need for March of Dimes donations to offset the cost of GG procurement meant that publicity was linked to economic survival. "I am certain the American people will contribute even more generously to the current March of Dimes, knowing that some of these funds will be used to try to stem the tide of polio epidemics this year," proclaimed Basil O'Connor on the ABC coast-to-coast TV network in January.[54] O'Connor introduced the nation's first polio immunization program by fusing the optimism of Cold War scientific discovery with military power; while the hydrogen bomb was presented as "the world's most devastating potential killer," GG was named "a potential protector of life and limb."[55] Like efforts to contain the perceived threat of communist expansion in Korea, the national program promised to contain polio at home.[56] Although O'Connor admitted that there would not be enough GG to protect all children, it would be made available in "severe epidemic areas."[57] For O'Connor, it was imperative that citizens associate the GG program with the NFIP as a means of fending off ARC encroachment and bolstering the March of Dimes.[58]

Beyond reinforcing the link between GG and the NFIP, officials worked to temper expectations about the national program. Irrespective of the debates over the untested household contacts method, most journalists were excited to report on a new medical intervention that might curb epidemics after decades of devastation wrought by polio. Out of enthusiasm, many writers praised the national program and its method. By way of this scheme, one writer chimed, "Every American family in which polio strikes this summer will be given a dose."[59] Although such coverage boosted public support, it seeded false hope and the potential for a backlash if GG failed to meet expectations.

To temper optimism, NFIP officials explained that it was important "not to expect too much from recent heartening research and continue to be sensible and alert."[60] They reminded readers that polio control was dependent not only on health professionals, but also on civilians "to assure community understanding and avoid unreasonable demands for GG."[61] The NFIP also presented GG as a stopgap solution in juxtaposition to encouraging news of a prototype vaccine developed by Dr. Jonas Salk.[62] "The objective has not yet been achieved," NFIP officials cautioned. "Widespread use of any polio vaccine must await proof of both safety and effectiveness."[63] By comparing GG with a vaccine, NFIP officials hoped to balance expectations and shift hopes onto a more promising medical intervention.

NFIP officials actively counteracted criticism of the national program and shielded their organization from misrepresentation. The consequences of an impending GG shortage roused some critics; one *New York Times* reporter asserted, "Protecting all 46 million American youngsters in the particularly susceptible one to eighteen age group through the many months of the polio season would require the GG from 200 million pints of blood—about two-hundred times the amount of gamma globulin actually available this summer."[64] NFIP officials assuaged such fears by explaining that efforts were in place to boost production and increase the number of doses. "About 60,000 additional average doses of gamma globulin each month will be available this summer," they promised.[65] Moreover, concerns about the effect of GG on the human immune system were swiftly addressed. Dr. Johannes Ipsen, superintendent of the Massachusetts Department of Public Health, delivered a cutting indictment of GG at a health conference.[66] Ipsen attacked what he saw as a misguided program that interfered with the acquisition of natural immunity. "This mass injection procedure," he declared, "will have no influence on the polio epidemic next year, and may even involve depriving several thousand of those persons who receive immune globulin of the benefit of naturally acquired immunity."[67] When newspapers carried this debate to readers, NFIP officials countered that "the scientific evidence to date does not support this opinion."[68] They also responded to concerns over GG efficacy. Dr. G. D. Cummings, a member of the ODM subcommittee on blood, expressed to the NFIP that he believed GG was not effective against polio and that "too much publicity" about the field trials cloaked the truth.[69] Ducas responded privately that while such opinions might be true, it was important that the NFIP "keep people informed," especially those "who had given us money with which to tackle the polio problem."[70] NFIP officials were forced into the uncomfortable position of defending the continued use of GG for polio control.

The NFIP mobilized its chapter volunteers and regional representatives to serve as logistics and publicity representatives. A training pamphlet was developed and released to help prepare "PR Men" to manage the program at a local level. The first objective was for representatives to build trust among doctors and health officers. "Don't be upset by initial coolness," the pamphlet counseled. "In no case reported to date has this attitude continued long." Once trust was established, representatives were encouraged to offer their assistance to local health professionals and provide clinic plans as requested.[71] "Do not propose this to the health officer. Instead, keep it in your pocket until he makes some such remark as, 'I wonder how the clinic should be set up,' or 'I wonder what supplies we'll need.'" NFIP representatives would then step in and offer a prepared solution.[72]

Besides operational assistance, representatives were asked to liaise with the press. "If you are identified as the polio news man early in the game," the pamphlet advised, "there will be a natural flow of information to you." Beyond local publicity, a further goal was to cultivate national awareness so that "folks in far-off places know what is going on." The benefits of a grassroots publicity program conducted by chapter volunteers would theoretically improve perceptions of the NFIP. "This increases NFIP prestige and public understanding of our work, which will be helpful come January," it concluded.[73]

Representatives were further required to write news stories, locate lollipop vendors for clinics, and stage flattering photographs. During these activities, they were advised to "stay in the story" by promoting NFIP affiliation. "If you are staging a special picture of the 200,000th national inoculation and your only hook is the presence of one of our volunteers in the picture, *do not allow this person to be shoved out*," the pamphlet explained. "If this happens, restage the picture. Many of the pictures that we take and service will be credit-less, but on the major pushes we have a right to identification."[74] NFIP publicists saw the generation of press releases and staged photographs as vital to helping the organization retain recognition for the national program and strengthen its annual fund-raising drive.

As doses of the ARC's GG and commercial GG were shipped by the ODM to public health departments across the United States before epidemic season, journalists welcomed the arrival of the national program with great fanfare.[75]

Time magazine designated summer "Gamma Globulin Season," while *Better Homes & Gardens* jibed, "Where Is That Much-Talked-About Gamma Globulin?"[76] Although the efficacy of the household contacts method was unproven, most citizens expressed little concern; instead, focus was trained on a collective hope that polio epidemics could be controlled.

The availability of GG created desperate scenes as parents clamored to protect their children. When polio struck a 4-H summer camp in Sutter County, California, thirty-six girls were identified as "household contacts" of a stricken bunkmate and rushed by their parents to family doctors for GG injections.[77] In the small town of Prairie Village, Kanas, eight cases of polio were reported; the ODM transported 117 doses of the blood fraction to the region and even though there was not enough serum to inject all "exposed children," the parents of "passed-over children" reportedly "took it gracefully."[78] In Chicago, Illinois, twenty-two children from the Kitty Kastle Day Nursery were allocated GG doses by the public health department after being exposed to the virus. Elsewhere in the city, a father attempted to bribe the board of health's chief clerk and president with $500 to obtain GG for his child, whom he believed was exposed to the virus. "It was a serious matter, attempting to undermine the integrity of a city employee," explained the board's president, "but his child had been exposed to polio and the father was desperate."[79] Many parents had come to believe in the national program and trusted in the power of GG.

Increasing public demand led to changes in the GG usage criterion in favor of mass prophylaxis. In June, Montgomery County, Alabama, was visited by a severe polio epidemic with eighty-five reported cases.[80] State health officer Dr. D. G. Gill was optimistic about inoculating a largely rural population, since his experience conducting surveys for syphilis and tuberculosis showed considerable popular support for public health programs.[81] On June 26, local doctors assented to Gill's ambitious plan for mass inoculation: "every child under 10 in the county" would be injected with GG over a four-day period.[82] The use of GG in Montgomery united military, philanthropic, and civilian agencies, as the state governor extended his public relations bureau for publicity and eight hundred local volunteers managed eighteen school-based clinics across the region.

Black and white families "flocked" to their nearest segregated school for GG injections. Emotions ran high, as clinic volunteers managed anxious parents and administered the blood fraction to children. Medical personnel from the nearby Maxwell and Gunter Air Force Bases helped "handle the unruly few" and gave "assurance to many of the timid."[83] "This," one health official stated, was "the finest example of cooperation I have ever seen." Although some parents were concerned by the potential risks of injection, they ultimately trusted the program. "I only want my child to have whatever can be given," reasoned one Alabama mother.[84] Nearly sixty-seven gallons of GG was administered to over thirty-two thousand Montgomery children. Some journalists praised the program as "a success in the sense that there has been a reduction both in the severity and the number of [polio] cases."[85] Other writers characterized the event as the "first flowering of a government-sponsored plan" to control epidemics.[86] Out of appreciation, adults volunteered to be "silent partners" by donating blood to the ARC.[87]

Between July 6 and August 7, 1953, Caldwell and Catawba Counties in North Carolina experienced severe outbreaks that led to mass prophylaxis.[88] In the New York counties of Chemung and Steuben, thirty-five thousand individuals received injections from one of twenty-three clinics managed by five thousand volunteers and 125 medical professionals.[89] Despite the original plan to use GG only for household contacts, state health officers exercised some agency in allocation within their regions. Although it was not known whether

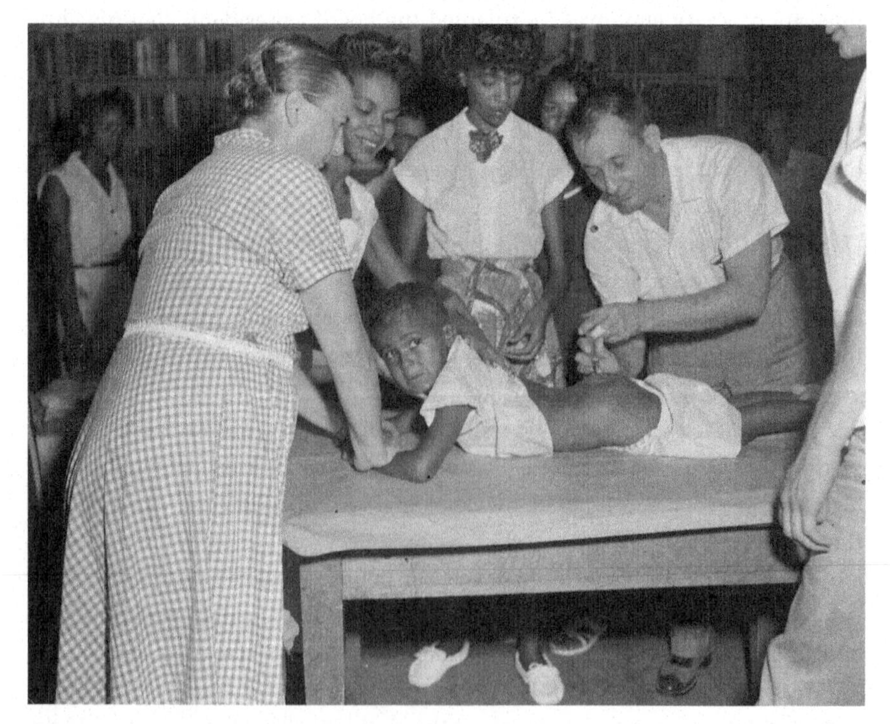

Figure 21 Five-year-old James Coachman receiving gamma globulin injection, 1953. Courtesy of the March of Dimes Foundation.

the national program reduced polio paralysis, many parents and health professionals believed it helped.

Protecting the National Program

Despite the initial favorable response to the national program, criticisms began to grow. The NFIP came under fire for the cost; in New York, the *Herald Tribune* revealed the total outlay of $221,000, which was "almost $13,000" to protect one child in two thousand from polio.[90] When reached for comment, Hammon admitted, "It's expensive, but it's the only thing we've got and if the public wants to pay for it, then they ought to get it."[91] The carefully selected proving grounds of 1951 and 1952 did not simulate the public health realities of the nation. It was evident that GG did not perform to the expected standard. Hammon privately admitted to the NFIP that the blood fraction was of limited public health value.[92] Based on projections, he estimated that in mass prophylaxis GG might prevent 1.6 cases of polio paralysis per 1,000 injections and in household contacts he expected the number to rise slightly to 2.2 cases per 1,000.[93] Observations from the field reinforced Hammon's appraisals. Seventeen days after six-year-old Peggy Robbins of North Carolina received her GG injection she began to suffer fever, leg weakness, and a stiff neck; she was diagnosed with polio and admitted to the hospital, where she remained for several weeks undergoing therapy for her "foot drop." Similarly, twenty-one days after fourteen-month-old Anita June Raby was injected with GG, she was diagnosed with polio.[94] It appeared that the supposed four-week window of polio protection, commencing the first week after injection, was not guaranteed.

Although they anticipated the lackluster performance for GG, NFIP officials grew concerned by critical scientific and public health reports. The World Health Organization (WHO), established after World War II as the global health arm of the United Nations, served as an important forum for public health research. In September 1953, the WHO convened a conference on polio in Rome, Italy, and invited heads of health charities and researchers, including Basil O'Connor, Dr. Jonas Salk, and Dr. Albert Sabin. At the conference, Yale University researcher and chair of the WHO Expert Committee on Poliomyelitis, Dr. John R. Paul, argued for the "condemnation of GG," based on evidence of its limited value at controlling polio.[95] Many attendees were startled by Paul's revelation and wondered what his committee's report meant for the continued application of GG.

In addition to the critical WHO report, NFIP officials were troubled by the forthcoming report by the USPHS' National Advisory Committee for the Evaluation of Gamma Globulin.[96] Beginning in April 1953, the USPHS had gathered data on GG allocations from twenty-one counties in thirteen states, one city, and one territory. Over 220,000 children received GG injections, and a total of 1,728,700 cc of GG was administered.[97] Usage criterion variations combined with an absence of controls meant that USPHS officials struggled to extract meaningful results.[98] Personnel shortages and the challenge of ensuring accurate diagnoses exacerbated reporting problems.[99] One physician acknowledged that 25 percent of "cases originally diagnosed as poliomyelitis and sent to hospitals [were] finally diagnosed as something else."[100] Despite data quality problems, USPHS officials believed their evidence showed that GG was ineffective at curbing paralytic polio. Out of courtesy, Dr. Alexander Langmuir sent the NFIP advance warning. "We find it exceedingly difficult," he explained, "to make a reasonably confident interpretation . . . that gamma globulin was truly beneficial."[101] One NFIP representative, who worked closely with the USPHS during the national program, spoke with Langmuir and reported back. "He first termed this data 'discouraging,'" the representative informed Ducas, "but later said flatly that it showed conclusively that use of gamma globulin in the family or immediate contact method failed; that use of gamma globulin in the mass inoculation method failed; that use of gamma globulin in both methods did not even modify the paralytic effects of the disease. He commented that 'Basil O'Connor told me I'd find that the family contact method would fail and I'll have to admit he was right.'"[102] Even though Langmuir observed that the national program helped "prevent panic" and cultivated "better public feeling," he admitted that the psychological benefits could not be considered "in a scientific study."[103] Based on this unofficial report, the national program was little more than an expensive charade.

News of the USPHS report posed complications for the NFIP. The timing was inopportune: the charity had already committed millions of dollars to expanding the GG program into 1954, and it was difficult to justify the investment unless it was "explained that the negative statement [about GG was] ill-founded."[104] The NFIP also needed to maintain faith in its research program; if Americans learned that GG was not useful, it might impinge on perceptions of future medical studies.[105] The charity was also poised to launch its March of Dimes drive, and negative publicity was sure to affect the generosity of donors.

After decades of fighting polio alongside the NFIP, some USPHS officials sympathized with the NFIP's situation and offered to mitigate the potential harm of the report. NFIP officials learned through private channels that the USPHS would try to "set the tone of the report so it will not be negative about GG" and attempt to postpone its release date, so that it would not "have an adverse effect" on the March of Dimes campaign.[106] One USPHS official even pledged to "see if the final report can be retained for background for future decisions rather than be made public."[107] Not all of these efforts were successful, but USPHS officials were able to withhold the report until after the fund-raising drive.[108] While this provided a temporary reprieve for the NFIP, the USPHS could not prevent publication of its report.

Convened by Surgeon General Dr. Leonard Scheele in January 1954, the USPHS's National Advisory Committee for the Evaluation of Gamma Globulin met to review the national program's performance.[109] Evidence showed that GG was often administered too late to provide protection, and even when it was given at the proper time its efficacy was limited.[110] Neither the mass prophylaxis nor household contacts method showed much value.[111] Obtaining GG through the ODM was found to be "cumbersome" because health officers had to demonstrate a high incidence of polio before the blood fraction was released.[112] Applications and paperwork caused delays and lost precious time. The committee further discovered polio antibody variations between the ARC's GG and commercial GG, suggesting that some doses of the blood fraction were more protective than others.[113] After the meeting, one committee member offered a cutting indictment: "Forget about gamma globulin for polio and turn it over to the states for use against measles and hepatitis."[114] Although admitting that their evaluation was not conclusive, committee members found little evidence to show that GG was a useful weapon against polio.

When the USPHS report was released in February 1954, it was met with a mixture of disappointment and astonishment. "Decision Reversed," judged *Time* magazine; "Find Gamma Globulin No Help in Polio" pronounced the *Chicago Tribune*.[115] Members of organizations that had participated in the program, such as the North Carolina Medical Society, were incredulous and counselled colleagues that GG was of "doubtful value for polio."[116] It was becoming clear that the hopes vested in GG had been misplaced.

Since it was difficult to halt the national GG program without admitting its shortcomings and offering an alternative, the NFIP fell back on its strongest weapon: marketing. Dorothy Ducas mobilized NFIP regional representatives to solicit optimistic testimonials from public health officers. According to Ducas, testimonials needed to be "well documented and backed up by first hand reports" in case the story was "questioned." Obtaining evidence in support of mass prophylaxis over household contacts was deemed to be imperative to the defense strategy. Presaging the benefits, Ducas reasoned, "I should like our story to read something like this: 'Nine out of fifteen county health officers who used GG in mass inoculations during 1953 believe the results were beneficial.' Of course, if more of them say 'no' than 'yes' we won't have a story, but based on what I know so far I don't think this will happen."[117] Regional representatives were asked to "make room" for this new mission and travel "to the health officers' cities" to gather the accounts.[118] As anticipated by Ducas, the collected testimonials trumpeted the efficacy of GG for mass prophylaxis. Most health officers were optimistic and "perfectly willing to be quoted." One Park County, Montana, health officer testified, "The mass inoculation cut the feet from under our epidemic. Polio disappeared almost immediately after completion of the project." Similar sentiment was expressed by the Saint Cloud, Minnesota, health officer, who reasoned that "there is no doubt in my mind as to the effectiveness of the mass inoculation use of gamma globulin. . . . [We] stopped the epidemic in its tracks."[119] Although such testimony could not be corroborated by the USPHS report, it could be used to defend the NFIP's investment in GG.

NFIP officials reeled from the frustrating situation. "If there had been any scientific grounds for opposing the program, you can bet I'd have opposed it to the end," remembered Dr. Thomas Rivers, "but there wasn't much I could do." The power and seemingly indisputable nature of scientific evidence generated by Hammon's controlled field trial stood as the basis to trust in the blood fraction and the national program. Attention to promoting Hammon's field trials created a wave of optimism that linked GG directly to the NFIP and its research program. "Wouldn't you have done the same?" Rivers continued. "Why do you suppose the people gave us money? They wanted us to fight polio. So we fought polio."[120] The NFIP mandate to fund medical research and show progress in the war against polio created a publicity monster that was difficult to slay.

To rationalize continuation of the national program into 1954, NFIP officials devised press releases and organized speaking engagements. Their strategy was to attack the failed household contacts method and defend Hammon's mass prophylaxis method. On February 23, NFIP publicists claimed that the poor performance of the household contacts method "was to be expected," since the NFIP recommendation was "not followed."[121] Special editorials were supplied to major newspapers and magazines; the May 24, 1954, edition of *Life* ran a NFIP editorial, titled "The Polio Season."[122] In the article, publicists spun the scientific debate about GG's efficacy as the "confusion of tongues" and maintained that GG had always been effective against polio, but poorly administered in the 1953 trials. NFIP medical director Dr. Hart Van Riper followed a similar tack at the American Orthopedic Association conference, extolling the benefits of GG for polio and claiming that nothing in the USPHS report "nullifies or modifies" earlier scientific evidence.[123] Likewise, O'Connor defended the use of GG during a CBS television interview on the popular *Longines Chronoscope* show.[124] When the interviewer asserted that the USPHS found "gamma globulin was not effective anymore," O'Connor interjected, "No, no, no, they never said that. The Public Health never said that. What happened was a committee set up by the Public Health Service made a report out of Atlanta that said the use of gamma globulin in family contacts was no good and we always said that; there was never any basis for believing it was any good."[125] O'Connor reminded viewers that the USPHS had not entirely condemned GG, since its own data failed to determine whether the blood fraction worked in mass prophylaxis. "The only scientific test made by Dr. Hammon in 1951 and 1952," O'Connor affirmed, "still proves that GG given at the proper time . . . and the proper kind of GG will protect." Rapping his hand on the studio desk, he declared that GG was available in 1954 and that it "should and it will be used."[126] By championing mass prophylaxis, NFIP officials hoped to deflect criticism and justify their sustained investment in the national program.

Hammon escaped most of the growing condemnation. Although his problematic earlier study laid the clinical foundation of GG for polio, his professional contributions were respected. He remained head of department and accepted consultancy work for the Armed Forces Epidemiology Board.[127] He expanded his laboratory facilities and recruited more teaching staff at the University of Pittsburgh. Hammon also launched a four-month research study in 1953 of subclinical polio infections among two hundred "normal families" and "native domestic servants" living at the Clark Field military base in the Philippines.[128] As his field work increased, he turned his attention to Japanese B encephalitis virus and its association with wild birds.[129] Although the NFIP could not be easily extricated from the deteriorating GG saga, Hammon escaped relatively unscathed.

Recognizing Hammon's publicity value, NFIP officials planned for his "reactivation" as a "protagonist" to the cause. He was asked to shelve his nuanced paper for the June 1954 American Medical Association conference for a ghostwritten "stronger statement." Reflecting on the opportunity to influence doctors as a form of religious conversion, NFIP officials remarked of the AMA conference, "It is hoped that much missionary work can be done there next week." Hammon was also asked to appear for "ten minutes on a nationwide radio broadcast" and on a NBC television show to extoll the virtues of GG.[130] Even his former field trial colleagues, Coriell and Stokes, were contacted to "consider the authorship of brief articles" promoting GG.[131] As bidden by his sponsors, Hammon issued a strong defense of GG for polio.[132] While he acknowledged its clinical limitations, he explained that GG worked well under specific conditions. "It has been and still is generally accepted," he reported, "that gamma globulin in an appropriate dose given under the proper circumstances to children in epidemic areas will afford protection to some for a limited period of time."[133] To uphold his convictions, Hammon embarked on a speaking tour, delivering papers at meetings of the American Pediatric Society and AMA.[134] He also published a special reassessment of GG in the *Journal of the American Medical Association*, claiming that GG was actually more potent than previously known and that its protective effect only diminished after six to eight weeks.[135] NFIP officials hoped that Hammon's tours, presentations, and publications would help transform GG from an impractical polio intervention to a stopgap success.

In addition to conscripting Hammon, NFIP officials issued pamphlets that emphasized "improvement" over the previous year and referenced "the most optimistic report possible on the 1953 use of GG."[136] In a special "Report to Physicians," NFIP assistant medical director Dr. Kenneth Landauer reiterated Hammon's claims that GG was "not only reaffirmed" through further study, but also "even greater than we formerly believed." He suggested that any failures in 1953 were the result of a "wave of false optimism," the "misapplication" of GG in household contacts, and the absence of statistical "controls" for proper evaluation. He also challenged the USPHS report by explaining that GG could be "of real value" if given "at the right time and place and in the right amount." Far from a defeat, Landauer framed GG as part of a "stopgap" solution until a vaccine was discovered.[137] Although NFIP officials attempted to overturn tarnished perceptions of the national program, declining faith in the blood fraction was difficult to counter. Only by turning Americans' attention to another intervention could they hope to save face.

Distracting from the Past, Welcoming the Future

By 1954, the landscape of polio control in America appeared divided and dysfunctional. The prospect of a killed-virus vaccine, developed by Hammon's adversary Dr. Jonas Salk, seemed imminent. Salk had tested variations of his prototype vaccine in small clinical trials in 1953, which determined that a series of spaced injections offered durable immunity against all three types of poliovirus.[138] Innovations in tissue culture techniques, including cultivating cells in Connaught Laboratories' nutrient medium 199, made the production of vast quantities of poliovirus for a vaccine possible.[139] Until the prototype vaccine could be further evaluated for safety and efficacy, GG remained the only option in the fight against polio. Although NFIP officials were discouraged by the national GG program and hoped to abandon it as quickly as possible, its continuation was essential to save face, honor pharmaceutical contracts, and provide a measure of hope.[140]

In 1954 the national GG program, unlike during its inaugural year, was received by parents and health officers with mixed feelings. Some were enthusiastic. As one scientist boasted, "If my children are in a polio epidemic area this summer, I hope they can get gamma globulin."[141] Over 1,900,000 doses of GG were available for polio control, which was enough to inoculate fifteen times the number of children than the year before.[142] Given the surplus, the ODM permitted health officers a "considerable latitude in administering" GG, but discouraged the household contacts method.[143] When outbreaks emerged, such as one in Key West, Florida, health departments commenced mass injections of expectant mothers and children.[144] Others were less optimistic. "I personally am not too enthusiastic about this use of GG," one health officer opined, "but I don't make the rules." Fifteen state health departments openly "expressed strong doubts" about the blood fraction and refused to use it at all.[145] The health department in Illinois declared that since GG was "ineffective," it "would not be permitted this year."[146] On July 1, 1954, NFIP funding for blood serum processing halted, and by October 1 the national program was ended.[147] Even though over $24 million of March of Dimes funds was spent by the NFIP to purchase and ship GG for the national program, it was difficult to continue in the face of public health objection.

Supporting the national GG program frustrated NFIP officials. Covering the cost of commercial blood fractionation and serum shipping drained an already dwindling coffer. Moreover, competition for public donations with the ARC complicated polio fund-raising and threatened financial contributions.[148] One advertisement appearing in the *New York Times* reminded readers that the ARC blood program provided the ODM with "some 7.5 million cc of gamma globulin to fight polio."[149] In addition, some NFIP supporters were dismayed by investment in a blood fraction that offered little protection against polio. NFIP officials recognized that only a fresh, exclusive, and well-promoted medical triumph could dislodge the ARC and restore Americans' faith in the March of Dimes.

A large and well-publicized test of Salk's prototype vaccine promised the necessary triumph. In December 1953, NFIP officials began preparations for an experiment enrolling over 1.8 million first-, second-, and third-grade schoolchildren.[150] Salk was initially opposed to the trial, as he believed his studies showed the vaccine was safe and effective, but he nevertheless agreed to the plan out of a need for external assessment.[151] Orchestrating a large vaccine trial was a strategic public relations move for the NFIP. If the vaccine was effective, such news would not only outshine the ARC and justify years of expenditure on medical research, but also help to make March of Dimes donors feel part of a success story. In *Today's Health,* Basil O'Connor admitted that "if an effective vaccine emerges from these studies . . . the layman—who furnished the original impetus for this effort—will have been instrumental in dealing the disease its final blow."[152] Perhaps most importantly, the trial would distract attention away from the embarrassing GG program. Even USPHS officials, such as Langmuir, encouraged the NFIP to undertake the trial to "offset the expected public reaction to the gamma globulin report and to get the public interested in the vaccine and its mind off gamma globulin."[153] After years of disappointment, NFIP officials knew that their donors needed renewed hope.

NFIP officials applied their experience with GG to the clinical trials of the first polio vaccine. Since Hammon's study showed that parents would volunteer their healthy children to an experiment that did not promise safety or efficacy, NFIP officials were confident that an enormous civilian medical study was achievable.[154] The use of publicity materials, such as films, radio announcements, and posters, were proven as vital recruitment tools. The legal consent form used in Hammon's study, far from being an aberration, became a model and was further refined for use in the vaccine trial.[155] Unlike Hammon's experiment, O'Connor decided that the vaccine field trial would be managed and assessed by a separate organization; he consequently funded the Vaccine Evaluation Center at the University of Michigan under the leadership of Salk's mentor, Dr. Thomas Francis Jr.[156] O'Connor hoped that external assessment would increase the scientific rigor and impartiality of the results. NFIP officials' involvement with GG had provided valuable lessons about selling science.

The prospect of two competing polio interventions, one tested and one under evaluation, complicated publicity and public education. Some NFIP officials feared that ignorance might lead some citizens to assume that GG was a vaccine or that the vaccine was no better than GG. To avoid misconceptions, NFIP publicists scrambled to update their literature.[157] In a 1954 educational leaflet, the NFIP emphasized that GG was "not a vaccine" and only protected for "five weeks."[158] They also attempted to restrict the use of GG during the vaccine study, so as not to "affect the validity" of field trial data.[159] Despite efforts, GG remained in widespread use, even at locations where the vaccine was being evaluated.[160] Although administering GG could undermine the accuracy of the vaccine trial, NFIP officials recognized that it was "exceedingly difficult for the organization to deny" people access to the blood fraction when there was no alternative.[161] Until a vaccine was tested, licensed, and available, GG was America's only approved polio intervention.

The NFIP bet its institutional credibility on the vaccine trial. Fortunately, the results were worthy of a spectacle. In anticipation, Basil O'Connor arranged for a public announcement of Francis's report on April 12, 1955, the tenth anniversary of Roosevelt's death, with a live broadcast to cinemas across the nation. Anticipation was high as Francis declared the vaccine "safe, effective, and potent."[162] Americans greeted the announcement with relief. As one Houston, Texas, resident recalled, "I was coming out of a store with my family when a truck screeched to a halt in front of us. A man was crying and shouting as he tossed down copies of a special edition of the newspaper. When everyone finally understood his words there was great joy—a giddiness that can only be experienced at the end of a time of terror."[163] The vaccine was quickly licensed by the federal government and shipments began immediately. The results of the vaccine field trial legitimized the efforts of the NFIP and helped America forget about its dubious fascination with GG for polio.[164]

GG slowly faded out of clinical use for the prevention of polio, but its application for other diseases continued. It was used for measles and later administered in concert with the measles vaccine to reduce adverse reactions.[165] It was also prescribed to pregnant women exposed to rubella (German measles). *Time* magazine reminded readers in 1964 that "if a woman who is pregnant, or thinks she may be, is exposed to German measles, she should get a shot of gamma globulin."[166] The blood fraction also remained important for hepatitis control and was regularly provided to international travelers.[167] The Peace Corps, a federal agency established under the Kennedy administration to provide assistance to developing nations, administered "massive shots of gamma globulin every four to six months" to guard its volunteers against hepatitis in "the unsanitary conditions where they work."[168] Demand for GG peaked in 1995, when its use for military personnel caused a shortage that affected international travelers. One *New York Times* article warned that "travelers leaving the United States for developing countries may have a hard time finding gamma globulin shots at the moment."[169] While the discovery of a hepatitis vaccine in the 1990s reduced dependence on GG, the blood fraction remains in clinical use.[170]

Although Hammon's GG study and the subsequent national program harbored risks and failed to exhibit a lasting solution to polio, they were a marketing success that showed scientists and the NFIP what civilians would tolerate during a crisis. Like the Tuskegee study of untreated syphilis, the GG trials revealed that the pursuit of knowledge, however unethical and limited in scientific value, can blind researchers and their sponsoring agencies. Exposure of healthy children to increased health risks, such as contagion and adverse reactions, were deemed acceptable to GG proponents because generating knowledge appeared more important. The GG program, fueled by prior momentum, personal and institutional agendas, and the need to control outbreaks, showed that the appearance of progress mattered. However, privileging appearance over safety and utility inadvertently undermined the foundations of scientific research, as well as an important precept of clinical medicine: *primum non nocere* (above all, do no harm).[171] Hammon's failure to properly recognize or acknowledge the shortcomings of his study permitted the generation of knowledge that was difficult to overturn. For the NFIP, selling the experiments and their aftermath became more important than the underlying science. The GG study and national program seemingly addressed a public health crisis and normalized applied research on an open population. Its unexpected legacy was to reveal the ways in which marketing could shape scientific research.

Notes

Introduction

1. L. Pearce Williams, "Chapter VIII: Passive Immunity to Poliomyelitis: The Field Testing of Gamma Globulin as a Prophylactic Agent," The Medical Research Program of the National Foundation for Infantile Paralysis, December 1956, March of Dimes Archives, White Plains, New York (henceforth MDA), 75.
2. "Polio Test Report Set for April 12," *New York Times*, March 23, 1955, 33; William L. Laurence, "Salk Polio Vaccine Proves Success; Millions Will Be Immunized Soon; City Schools Begin Shots April 25," *New York Times*, April 13, 1955, 1; Jeffrey Kluger, *Splendid Solution: Jonas Salk and the Conquest of Polio* (New York: G. P. Putnam's Sons, 2004), 294; John R. Paul, *A History of Poliomyelitis* (New Haven: Yale University Press, 1971), 432.
3. Michael B. A. Oldstone, *Viruses, Plagues, and History* (New York: Oxford University Press, 1998), 104.
4. David M. Oshinsky, *Polio: An American Story: The Crusade That Mobilized the Nation against the 20th Century's Most Feared Disease* (Oxford: Oxford University Press, 2005), 8.
5. "Medicine: Polio at Work," *Time*, September 20, 1948, http://www.time.com/time/magazine/article/0,9171,799172,00.html, last viewed July 12, 2011.
6. Oshinsky, *Polio: An American Story*, 8.
7. Paul, *A History of Poliomyelitis*, 7.
8. Frederick C. Robbins, "The History of Polio Vaccine Development," in *Vaccines*, 4th ed., ed. Stanley A. Plotkin and Walter A. Orenstein (Philadelphia: Elsevier, 2004); Oldstone, *Viruses, Plagues, and History*, 104.
9. Naomi Rogers, *Dirt and Disease: Polio before FDR* (New Brunswick, N.J.: Rutgers University Press, 1992).
10. "Medicine: Polio Season," *Time*, September 8, 1941, http://www.time.com/time/magazine/article/0,9171,849501,00.html, last viewed March 3, 2011.
11. Oshinsky, *Polio: An American Story*, 70.
12. Nancy Tomes, *The Gospel of Germs: Men, Women, and the Microbe in American Life* (Cambridge: Cambridge University Press, 1998).
13. Richard Carter, *The Gentle Legions* (New York: Doubleday & Company, 1961), 14. "Flies, Food, and Poliomyelitis," *JAMA* 128, 6 (1945): 442–443; "Dimes Are Sought in Paralysis Drive," *New York Times*, January 24, 1938, 23; Paul Strathern, *A Brief History of Medicine: From Hippocrates to Gene Therapy* (London: Constable & Robinson, 2005), 364; A. E. Casey, W. I. Fishbein, and H. N. Bundesen, "Transmission of Poliomyelitis by Patient to Patient Contact," *JAMA* 129, 17 (1945): 1141–1145.
14. Rogers, *Dirt and Disease*, 29–32.
15. Thomas M. Daniel and Frederick C. Robbins, eds., *Polio* (Rochester, N.Y.: University of Rochester Press, 1997); Hugh Gregory Gallagher, *FDR's Splendid Deception* (New York: Dodd, Mead, 1985), 2.
16. Heather Green Wooten, *The Polio Years in Texas: Battling a Terrifying Unknown* (College Station: Texas A&M University Press, 2009), 91; Marc Shell, *Polio and Its Aftermath: The Paralysis of Culture* (Cambridge, Mass.: Harvard University Press, 2005), 112.

17. F. Martin Harmon, *The Warm Springs Story: Legacy and Legend* (Macon, Ga.: Mercer University Press, 2014).
18. Oshinsky, *Polio: An American Story*, 24–28; Conrad Black, *Franklin Delano Roosevelt: Champion of Freedom* (New York: Public Affairs, 2003), 138–140; Victor Cohn, *Four Billion Dimes* (Minneapolis: Minneapolis Star and Tribune, 1955), 9.
19. Oshinsky, *Polio: An American Story*, 38–39; Black, *Franklin Delano Roosevelt*, 169–175; Rogers, *Dirt and Disease*, 168; Susan Mechele Ward, "Rhetorically Constructing a 'Cure': FDR's Dynamic Spectacle of Normalcy" (PhD diss., Regent University, 2005), 64; Jane S. Smith, *Patenting the Sun: Polio and the Salk Vaccine* (New York: William Morrow, 1990), 57–61.
20. Gallagher, *FDR's Splendid Deception*, chapters 5 and 6.
21. William H. Helfand, Jan Lazarus, and Paul Theerman, "'. . . So That Others May Walk': The March of Dimes," *AJPH* 91, 8 (August 2001): 1190; Heather Green Wooten, "The Polio Years in Harris and Galveston Counties, 1930–1962" (PhD diss., University of Texas at Galveston, 2006), 67.
22. Cohn, *Four Billion Dimes*, 55; Kathryn Black, *In the Shadow of Polio: A Personal and Social History* (Cambridge: Perseus Publishing, 1996), 25.
23. Paul V. Dutton, *Differential Diagnosis: A Comparative History of Health Care Problems and Solutions in the United States and France* (Ithaca, N.Y.: Cornell University Press, 2007), chapter 1; Sills, *The Volunteers: Means and Ends in a National Organization*. Glencoe, Ill.: Free Press, 1957, 170.
24. David W. Rose, *Images of America: March of Dimes* (Charleston, S.C.: Arcadia Publishing, 2003), 2.
25. Smith, *Patenting the Sun*, 52–53.
26. Ibid., 61.
27. Ibid., 67.
28. "Organizational Chart, 1949," Organizational Charts, NFIP, 1949–1953, S 10: Incorporation, Box 10, Medical Program Records (henceforth MPR), MDA.
29. Sills, *The Volunteers*, 23–25, 39.
30. Smith, *Patenting the Sun*, 65.
31. Stephen E. Mawdsley, "'Dancing on Eggs': Charles H. Bynum, Racial Politics, and the National Foundation for Infantile Paralysis, 1938–1954," *Bulletin of the History of Medicine* 84, 2 (Summer 2010): 217–247.
32. Charlene Pugleasa in *A Paralyzing Fear: The Triumph Over Polio in America*, ed. Nina G. Seavey, Jane S. Smith, and Paul Wagner (New York: TV Books, 1998), 121.
33. Oshinsky, *Polio: An American Story*, 65, 73–76, 239–40.
34. Mark O'Brien in Seavey et al., *A Paralyzing Fear*, 106.
35. Joan Elizabeth Morris, *Polio & Me, Now & Then* (Bloomington, Ind.: AuthorHouse, 2004), 3.
36. Charlene Pugleasa in Seavey et al., *A Paralyzing Fear*, 124.
37. Daniel J. Wilson, *Living with Polio: The Epidemic and Its Survivors* (Chicago: University of Chicago Press, 2005); Oshinsky, *Polio: An American Story*, 61–64.
38. Charlene Pugleasa in Seavey et al., *A Paralyzing Fear*, 127.
39. Victor Cohn, *Sister Kenny: The Woman Who Challenged the Doctors* (Minneapolis: University of Minnesota Press, 1975); Naomi Rogers, *Polio Wars: Sister Kenny and the Golden Age of American Medicine* (New York: Oxford University Press, 2014), Stephen E. Mawdsley Review of *Polio Wars*, *BHM* 89, 3 (Fall 2015): 623–624.
40. John Affleldt in Seavey et al., *A Paralyzing Fear*, 136.

41. F. Martin Harmon, *The Warm Springs Story*; Mawdsley, "'Dancing on Eggs'"; Naomi Rogers, "Race and the Politics of Polio: Warm Springs, Tuskegee, and the March of Dimes," *AJPH* 97, 4 (May 2007): 784–795; Edith P. Chappell and John F. Hume, "A Black Oasis: Tuskegee's Fight against Infantile Paralysis, 1941–1975" (Unpublished manuscript, March of Dimes Birth Defects Foundation, 1987).
42. Clara Yelder in Seavey et al., *A Paralyzing Fear*, 155.
43. Robert C. Huse, *Getting There: Growing Up with Polio in the 30's* (Bloomington, Ind.: 1stBooks, 2002), 108, 111.
44. Mark O'Brien in Seavey et al., *A Paralyzing Fear*, 108.
45. Anthony J. Badger, *The New Deal: The Depression Years, 1933–1940* (London: Macmillan Education, 1989);Ira Katznelson, *Fear Itself: The New Deal and the Origins of Our Time* (New York: W. W. Norton, 2013).
46. Morris, *Polio & Me, Now & Then*, 4.
47. Richard L. Bruno, *The Polio Paradox: What You Need to Know* (New York: Warner Books, 2002).
48. Mark O'Brien in Seavey et al., *A Paralyzing Fear*, 109.
49. Huse, *Getting There*, 140.
50. Gallagher, *FDR's Splendid Deception*.
51. Shell, *Polio and Its Aftermath*, chapter 5; Wilson, *Living with Polio*, chapters 6 and 7.
52. Morris, *Polio & Me, Now & Then*, 10.
53. Ibid., 11.
54. Shell, *Polio and Its Aftermath*, 140.
55. Simi Linton, *Claiming Disability* (New York: New York University Press, 1998), chapters 3 and 4; Lennard J. Davis, ed., *The Disability Studies Reader* (New York: Routledge, 1997), chapters 8 and 11.
56. Richard Aldrich in Seavey et al., *A Paralyzing Fear*, 113.
57. Charlene Pugleasa, ibid., 128.
58. Shell, *Polio and Its Aftermath*; Phillip Roth, *Nemesis* (London: Jonathan Cape, 2010).
59. Black, *In the Shadow of Polio*, 160–161.
60. Shell, *Polio and Its Aftermath*, 70–71.
61. Charlene Pugleasa in Seavey et al., *A Paralyzing Fear*, 131.
62. Blair L. M. Kelley, *Right to Ride: Streetcar Boycotts and African American Citizenship in the Era of Plessy v. Ferguson* (Chapel Hill: University of North Carolina Press, 2010).
63. Edward H. Beardsley, "Desegregating Southern Medicine, 1945–1970," *International Social Science Review* 71, 1 (2001): 37–54; Darlene Clark Hine, "Black Professionals and Race Consciousness: Origins of the Civil Rights Movement, 1890–1950," *Journal of American History* 89, 4 (March 2003).
64. James Jones, *Bad Blood: The Tuskegee Syphilis Experiment* (New York: The Free Press, 1981); Todd Savitt, *Race and Medicine in Nineteenth- and Early-Twentieth-Century America* (Kent, Ohio: Kent State University Press, 2006).
65. Francis to Weaver, December 10, 1946, S 14: Polio, Box 15, MPR, MDA.
66. "Paralysis Center Set Up for Negroes," *New York Times*, May 22, 1939, 15.
67. "Polio Cases in U.S. Up 71% This Year," *New York Times*, August 2, 1946, 11.
68. Naomi Rogers, "Polio Can Be Conquered: Science and Health in the United States from Polio Polly to Jonas Salk," in *Silent Victories: The History and Practice of Public Health in Twentieth-Century America*, ed. John W. Ward and Christian Warren (Oxford: Oxford University Press, 2007).

69. Rose, *Images of America*, 16; Susan Richards Shreve, *Warm Springs: Traces of a Childhood at FDR's Polio Haven* (Boston: Houghton Mifflin Company, 2007), 182.
70. Oshinsky, *Polio: An American Story*, 68; Smith, *Patenting the Sun*, 75.
71. Oshinsky, *Polio: An American Story*, 68.
72. Charlene Pugleasa in Seavey et al., *A Paralyzing Fear*, 123.
73. Oshinsky, *Polio: An American Story*, 68.
74. Daniel J. Wilson, "Basil O'Connor, the National Foundation for Infantile Paralysis, and the Reorganization of Polio Research in the United States, 1935–41," *Journal of the History of Medicine and Allied Sciences*. Published online, doi:10.1093/jhmas/jru003 (March 2014): 1–31.
75. Daniel J. Wilson, Polio (Westport, Conn.: Greenwood Publishing Group, 2009), 50.
76. Wilson, "Basil O'Connor, the National Foundation," 17–20; Oshinsky, *Polio: An American Story*, 112, 119, 154.
77. Paul, *A History of Poliomyelitis*, 311.
78. Karl E. Peace and Ding-Geng (Din) Chen, *Clinical Trial Methodology* (New York: Chapman & Hall/CRC Biostatistics Series, 2011), 1, 3.
79. Oonagh Corrigan and Richard Tutton, "What's in a Name? Subjects, Volunteers, Participants and Activists in Clinical Research," *Clinical Ethics* 1 (2006): 101–104.
80. Margaret A. Keller and E. Richard Stiehm, "Passive Immunity in Prevention and Treatment of Infectious Diseases," *Clinical Microbiology Reviews* 13, 4 (October 2000): 602–614.
81. Peter Keating and Alberto Camrosio, eds., *Cancer on Trial: Oncology as a New Style of Practice* (Chicago: University of Chicago Press, 2012).
82. Oshinsky, *Polio: An American Story*; Smith, *Patenting the Sun*; Kluger, *Splendid Solution*.
83. Susan E. Lederer, *Subjected to Science: Human Experimentation in America before the Second World War* (Baltimore: Johns Hopkins University Press, 1997); Allan M. Brandt, "Polio, Politics, Publicity, and Duplicity: Ethical Aspects in the Development of the Salk Vaccine," *International Journal of Health Services* 8, 2 (1978): 257–270.

Chapter 1 — Forging Momentum

1. "Meeting at Palmer House, Chicago, Illinois," August 25, 1945, "Gamma Globulin Study and Meeting with NFIP, 1945–1946," Box 50, Thomas Francis Papers, Bentley Historical Library, Ann Arbor, Michigan (henceforth BHL), 2.
2. "Study of Gamma Globulin Prophylaxis of Poliomyelitis, Freeport, Illinois," August 29, 1945, GG Program, 1944–1951, S 14: Polio, Box 12, Medical Program Records (henceforth MPR), March of Dimes Archives, White Plains, New York (henceforth MDA).
3. George B. Bader, "The Intramuscular Injection of Adult Whole Blood as Prophylactic against Measles: With a Report on the Literature," *JAMA* 93 (August 1929): 668–670; Charles F. McKhann, "The Prevention and Modification of Measles," *JAMA* 109 (December 1937): 2034–2038; George H. Weaver and T. T. Crooks, "The Use of Convalescent Serum in the Prophylaxis of Measles," *JAMA* 82 (January 1924): 204–206; "County-Wide Use of Immune Globulin in the Modification and Prevention of Measles," *JAMA* 106 (May 1936): 1781–1783; F. M. Meader, "Scarlet Fever Prophylaxis: Use of Blood Serum from Persons Who Have Recovered from Scarlet Fever," *JAMA* 94 (March 1930): 622–625; Wallace Sako, P. F. Dwan, and E. S. Platou, "Sulfanilamide and Serum in the Treatment and Prophylaxis of Scarlet Fever," *JAMA* 111 (September 1938): 995–997.

4. Dr. Charles Armstrong et al., *The Harvey Lectures—1940–1941* (Lancaster, Pa.: Science Press Printing Co., 1941); Edward A. Beeman, "Charles Armstrong, M.D.: A Biography," Office of History, National Institutes of Health, 2007, 214–215, https://history.nih.gov/research/downloads/ArmstrongBiography.pdf, last viewed December 19, 2008.
5. Arne Tiselius, "Electrophoresis of Serum Globulin, II. Electrophoretic Analysis of Normal and Immune Sera," *Biochemical Journal* 31, 9 (September 1937): 1464–1477; Charles A. Janeway, *The Gamma Globulins* (Boston: Little, Brown, 1966), 2.
6. Arne Tiselius and Elvin A. Kabat, "An Electrophoretic Study of Immune Sera and Purified Antibody Preparations," *Journal of Experimental Medicine* 69, 1 (January 1939): 119–131.
7. Angela Creager, "'What Blood Told Dr Cohn': World War II, Plasma Fractionation, and the Growth of Human Blood Research," *Studies in History and Philosophy of Biological and Biomedical Sciences* 30, 3 (1999): 379; Douglas M. Surgenor, *Edwin J. Cohn and the Development of Protein Chemistry: With a Detailed Account of His Work on the Fractionation of Blood during and after World War II* (Cambridge, Mass.: Harvard University Press, 2002).
8. Creager, "'What Blood Told Dr Cohn,'" 379.
9. Ducas to Weaver, July 23, 1951, "Background Material," S 14: Polio, Box 12, MPR, MDA; Creager, "'What Blood Told Dr Cohn,'" 330.
10. "Dr. Edwin J. Cohn, Blood Specialist," *New York Times*, October 3, 1953, 17.
11. Creager, "'What Blood Told Dr Cohn,'" 380.
12. Douglas Starr, *Blood: An Epic History of Medicine and Commerce* (New York: Alfred A. Knopf, 1998), 101–120.
13. L. Pearce Williams, "Chapter VIII: Passive Immunity to Poliomyelitis," The Medical Research Program of the National Foundation for Infantile Paralysis, December 1956, Revised Draft, MDA, 927.
14. "Dr. Joseph Stokes, Jr.," in "Background Story," 1952, S 3: GG FT, Box 3, Surveys and Studies Records (henceforth SSR), MDA, 11; Dr. Horace L. Hode, "Presentation of the John Howland Medal and Award to Dr. Joseph Stokes, Jr.," *American Journal of Diseases of Children* 104, 5 (1962): 440–442.
15. "Dr. Joseph Stokes, Jr.," in "Background Story," 1952, S 3: GG FT, Box 3, SSR, MDA, 11.
16. Ibid.
17. The Children's Hospital of Philadelphia, "Timeline," http://www.research.chop.edu/about/timeline/, last viewed March 2, 2010.
18. J. Stokes Jr. and J. R. Neefe, "The Prevention and Attenuation of Infectious Hepatitis by Gamma Globulin," *JAMA* 127 (January 1945): 144; "Medicine: Globulin v. Jaundice," *Time*, February 5, 1945, http://www.time.com/time/magazine/article/0,9171,797060,00.html, last viewed December 2, 2010; Jennifer A. Cuthbert, "Gamma Globulin (Passive Immunoprophylaxis)," *Clinical Microbiology Reviews* 14, 1 (January 2001): 38–58.
19. J. Stokes Jr., E. P. Maris, and S. S. Gellis, "Chemical, Clinical, and Immunological Studies on the Products of Human Plasma Fractionation. XI. The Use of Concentrated Normal Human Serum Gamma Globulin (Human Immune Serum Globulin) in the Prophylaxis and Treatment of Measles," *Journal of Clinical Investigation* 23, 4 (July 1944): 531–540; "Medicine: Blood v. Measles," *Time*, June 5, 1944, http://www.time.com/time/magazine/article/0,9171,778195,00.html, last viewed June 10, 2010.

20. "Medicine: Infantile Paralysis Vaccine," *Time*, January 1, 1934, http://www.time.com/time/magazine/article/0,9171,746680,00.html, last viewed May 2, 2010.
21. S. D. Kramer, K. H. Hendrie, W. L. Aycock, "Rise in Temperature Preceding the Appearance of Symptoms in Experimental Poliomyelitis," *Journal of Experimental Medicine* 51 (1930): 933–941; S. D. Kramer and M. Schaeper, "Experimental Poliomyelitis: Active Immunization with Neutralized Mixtures of Virus and Serum," *Proceedings of the Society for Experimental Biology and Medicine* 31 (1933): 409–411; S. D. Kramer and L. H. Grossman, "Active Immunization against Poliomyelitis. A Comparative Study," *Journal of Immunology* 31 (1936): 183–189; S. D. Kramer, B. Hoskwith, and L. H. Grossman, "Detection of the Virus of Poliomyelitis in the Nose and Throat and Gastro-Intestinal Tract of Human Beings and Monkeys," *Journal of Experimental Medicine* 69 (1939): 49–67; S. D. Kramer and H. A. Geer, "The Development of Active Immunity in Swiss Mice with Infective and Non-Infective Suspensions of Poliomyelitis Virus," *Journal of Immunology* 50 (1945): 275–281; "Meeting on Gamma Globulin, Waldorf Astoria, New York City," February 3, 1950, S 3: Conferences, Box 5, Conferences and Meetings Records (henceforth CMR), MDA, 26; L. Pearce Williams, "Chapter VIII: Passive Immunity," Preliminary Draft, 1956, S 1: Passive Immunity, Box 1, History of the NFIP Records (henceforth HNR), MDA, 6.
22. "Dr. Don Gudakunst, Paralysis Expert," *New York Times*, January 21, 1946, 20; "Donald W. Gudakunst," *Journal of Nervous and Mental Disease* 103, 3 (March 1946); "From the Archives," University of Michigan School of Public Health, *Findings Magazine* 23, 1 (Fall/Winter 2007).
23. Gudakunst to O'Connor, July 13, 1944, S 14: Polio, Box 12, MPR, MDA.
24. Williams, "Chapter VIII: Passive Immunity," Revised Draft, 928–929.
25. "The Present Status of the Usefulness of Convalescent or Human Serum in the Prevention and Treatment of Acute Poliomyelitis and the Experimental Infection," May 27, 1944, S 14: Polio, Box 12, MPR, MDA.
26. J. Stokes Jr., I. J. Wolman, H. C. Carpenter, and J. Margows, "Prophylactic Use of Parents' Whole Blood in Anterior Poliomyelitis; Philadelphia Epidemic of 1932," *American Journal of Diseases of Children* 50 (1935): 581–595; J. F. Kessel, A. S. Hoyt, and R. T. Fisk, "Use of Serum and the Routine and Experimental Laboratory Findings in the 1934 Poliomyelitis Epidemic," *AJPH* 24 (1934): 1215–1223; J. Stokes Jr. et al., "Chemical, Clinical, and Immunological Studies . . . XI. Use of Concentrated Normal Human Serum Gamma Globulin": 531–540.
27. Stokes to Cohn and Gudakunst, May 30, 1944, S 14: Polio, Box 12, MPR, MDA.
28. Starr, *Blood*, 95–110.
29. Stokes to Cohn and Gudakunst, May 30, 1944, S 14: Polio, Box 12, MPR, MDA.
30. Dochez to Kramer, July 14, 1944, S 14: ibid.; Cohn to Peet, August 12, 1944, S 14: ibid.; Theodore E. Woodward, *The Armed Forces Epidemiological Board: Its First Fifty Years* (Falls Church, Va.: Office of the Surgeon General, 1990), 46.
31. Theodore E. Woodward, "John H. Dingle," *Transactions of the American Clinical and Climatological Association* 85 (1974): xxxiii–xxxiv.
32. Dingle to Maxcy, July 20, 1944, S 14: Polio, Box 12, MPR, MDA.
33. Kramer to Gudakunst, June 14, 1944, S 14: ibid.
34. "Mich. Health Dept. X-Rays 11,735 at Fairs," *Billboard Magazine*, December 7, 1946, 67; DeKleine to Gudakunst, June 23, 1944, S 14: Polio, Box 12, MPR, MDA.
35. "Dr. DeKleine Retires as Head of Red Cross," *JAMA* 117, 23 (1941): 1995; J.A.K., "Blood Plasma Reservoir," *JAMA* (May 1941): 138; Starr, *Blood*, 95; Spencie Love, *One Blood* (Chapel Hill: University of North Carolina Press, 1996), 207–209.

36. Williams, "Chapter VIII: Passive Immunity," Preliminary Draft, 1956, S 1: Passive Immunity, Box 1, HNR, MDA, 9.
37. DeKleine to Gudakunst, June 26, 1944, S 14: Polio, Box 12, MPR, MDA.
38. DeKleine to Gudakunst, June 23, 1944, S 14: ibid.; DeKleine to Gudakunst, June 26, 1944, S 14: ibid.; DeKleine to Gudakunst, June 28, 1944, S 14: ibid.
39. Gudakunst to O'Connor, July 13, 1944, S 14: ibid., 1.
40. Richard Carter, *The Gentle Legions* (New York: Doubleday & Company, 1961), 60.
41. Gudakunst to O'Connor, July 13, 1944, S 14: Polio, Box 12, MPR, MDA, 1–2.
42. Special Committee Meeting, Resolution, July 19, 1944, S 14: Polio, Box 12, MPR, MDA.
43. Kramer, "Suggestions," July 19, 1944, S 14: ibid.
44. Gudakunst to Files, August 8, 1944, S 14: ibid.; Williams, "Chapter VIII: Passive Immunity," Preliminary Draft, 1956, S 1: Passive Immunity, Box 1, HNR, MDA, 8.
45. NFIP, "Proposal for a Field Study of the Value of Gamma Globulin as a Prophylactic Agent in Poliomyelitis," July 25, 1944, S 14: Polio, Box 12, MPR, MDA, 3–4.
46. Williams, "Chapter VIII: Passive Immunity," Preliminary Draft, 1956, S 1: Passive Immunity, Box 1, HNR, MDA, 10.
47. Kramer, "Suggestions," July 19, 1944, S 14: Polio, Box 12, MPR, MDA.
48. Gudakunst to O'Connor, July 13, 1944, S 14: ibid., 1.
49. Gudakunst to O'Connor, July 24, 1944, S 14: Polio, Box 12, MPR, MDA.
50. Gudakunst to DeKleine, August 3, 1944, S 14: ibid.; O'Connor to DeKleine, August 25, 1944, S 14: ibid.
51. Gudakunst to O'Connor, August 4, 1944, S 14: ibid.
52. Gudakunst to Files, August 8, 1944, S 14: ibid.
53. News from the Field, "Recurrence of Keratoconjunctivitis in Detroit," *AJPH* (February 1944): 210.
54. Gudakunst to Files, August 8, 1944, S 14: Polio, Box 12, MPR, MDA.
55. O'Connor to DeKleine, August 25, 1944, S 14: ibid.
56. Wilson to O'Connor, August 23, 1944, S 14: ibid.
57. DeKleine to O'Connor, August 17, 1944, S 14: ibid.
58. Wilson to O'Connor, August 23, 1944, S 14: ibid.
59. O'Connor to DeKleine, August 25, 1944, S 14: ibid.
60. Julia Irwin, *Making the World Safe* (New York: Oxford University Press, 2013), 205.
61. DeKleine to O'Connor, September 1, 1944, S 14: Polio, Box 12, MPR, MDA; Cohn to Kramer, December 16, 1944, S 14: ibid.; Kramer to Gudakunst, October 17, 1944, S 14: ibid.; Gudakunst to Cohn, October 21, 1944, S 14: ibid.
62. Kramer to Cohn, October 17, 1944, S 14: ibid.
63. "Michigan Mirror," *Cass City Chronicle*, September 21, 1945, 6.
64. "Study of Gamma Globulin Prophylaxis of Poliomyelitis, Freeport, Illinois," August 29, 1945, S 14: Polio, Box 12, MPR, MDA.
65. "Meeting at Palmer House, Chicago, Illinois," August 25, 1945, "Gamma Globulin Study and Meeting with NFIP, 1945–1946," Box 50, Thomas Francis Papers, BHL, Ann Arbor, Michigan; "Meeting on Gamma Globulin," February 3, 1950, S 3: Conferences, Box 5, CMR, MDA, 115–118; "Books Received," *AJPH* (September 1946): 1076.
66. "News from the Field," *AJPH* 33 (November 1943), 1383; "Army Experts to Tell Work of Medical Corps," *Chicago Tribune*, November 26, 1944, 4; "Polio Precautions Listed By State Health Board," *The Daily Illini*, August 10, 1946, 2.
67. "Study of Gamma Globulin Prophylaxis of Poliomyelitis, Freeport, Illinois," August 29, 1945, S 14: Polio, Box 12, MPR, MDA. Stokes, Report, 1944, St65p, "Research—Globulin," Joseph Stokes, Jr., Papers, American Philosophical Society.

68. "Meeting on Gamma Globulin," February 3, 1950, S 3: Conferences, Box 5, CMR, MDA, 115.
69. "Study, Freeport, Illinois," August 29, 1945, S 14: Polio, Box 12, MPR, MDA.
70. "Meeting at Palmer House, Chicago, Illinois," August 25, 1945, "Gamma Globulin Study and Meeting with NFIP, 1945–1946," Box 50, Thomas Francis Papers, BHL, 2.
71. "Meeting on Gamma Globulin," February 3, 1950, S 3: Conferences, Box 5, CMR, MDA, 116.
72. "Study, Freeport, Illinois," August 29, 1945, S 14: Polio, Box 12, MPR, MDA, 2; "Meeting at Palmer House, Chicago, Illinois," August 25, 1945, "Gamma Globulin Study and Meeting with NFIP, 1945–1946," Box 50, Thomas Francis Papers, BHL, 2–9.
73. "Meeting on Gamma Globulin," February 3, 1950, S 3: Conferences, Box 5, CMR, MDA, 118–119.
74. Irwin, *Making the World Safe*, 207.
75. Elizabeth W. Etheridge, *Sentinel for Health: A History of the Centers for Disease Control* (Berkeley: University of California Press, 1992).
76. "Dr. Don Gudakunst, Paralysis Expert," *New York Times*, January 21, 1946, 20.
77. William C. Reeves, "Arbovirologist and Professor, UC Berkeley School of Public Health," interview by Sally Smith Hughes, Regional Oral History Office, Bancroft Library, University of California, Berkeley, 1993, 66–68; David M. Oshinsky, *Polio: An American Story: The Crusade That Mobilized the Nation against the 20th Century's Most Feared Disease* (Oxford: Oxford University Press, 2005), 112–113.
78. A. Bloxsom, "Use of Immune Serum Globulin (Human) as Prophylaxis against Poliomyelitis," *Texas State Journal of Medicine* 75 (1949): 468–470; Heather Green Wooten, *The Polio Years in Texas: Battling a Terrifying Unknown* (College Station: Texas A&M University Press, 2009), 86.
79. "Epidemic Hits Houston," *San Antonia Light*, June 18, 1948, 2-A.
80. "Meeting on Gamma Globulin," February 3, 1950, S 3: Conferences, Box 5, CMR, MDA, 71–72. See also Bloxsom, "Use of Immune Serum Globulin (Human) as Prophylaxis against Poliomyelitis," *Texas State Journal of Medicine* 75 (1949): 468–470.
81. Charles R. Rinaldo, "Passive Immunization against Poliomyelitis: The Hammon Gamma Globulin Field Trials, 1951–1953," *AJPH* 95, 5 (May 2005): 791.
82. "Polio Expert's Career Started in a Jungle," *Houston Chronicle*, July 2, 1952, D-10.
83. Reeves, interview, 45–46. See also Rinaldo, "Passive Immunization," 791; "Background Story," 1952, S 3: GG FT, Box 3, SSR, MDA, 9.
84. Richard Carter, *Breakthrough: The Saga of Jonas Salk* (New York: Pocket Books, 1967), 100.
85. Michael Kasongo, *History of the Methodist Church in the Central Congo* (Lanham, Md.: University Press of America, 1998), chapter 2; Adam Hochschild, *King Leopold's Ghost: A Story of Greed, Terror and Heroism* (Basingstoke, UK: Pan Macmillan, 2012); Thomas Pakenham, *The Scramble for Africa: White Man's Conquest of the Dark Continent, 1879 to 1912* (New York: Avon Books, 1992); "Belgian Congo," *Encyclopedia Britannica*.
86. Reeves, interview, 45–46.
87. Glenn Fowler, "Dr. William Hammon Dies at 85: A Pioneer in Fight against Polio," *New York Times*, September 23, 1989, Obituaries.
88. "Polio Expert's Career Started in a Jungle," *Houston Chronicle*, D-10.
89. Reeves, interview, 45–46.

90. "Polio Expert's Career Started in a Jungle," *Houston Chronicle*, D-10.
91. Paul V. Dutton, *Differential Diagnosis: A Comparative History of Health Care Problems and Solutions in the United States and France* (Ithaca, N.Y.: Cornell University Press, 2007), 71; Paul Starr, *The Social Transformation of American Medicine* (New York: Basic Books, 1982), 270–279.
92. Reeves, interview, 45–46.
93. "Biographical Sketches of Dr. William McD. Hammon and Assistants," "Background Story," 1951, S 3: GG FT, Box 3, SSR, MDA, 6–7.
94. Reeves, interview, 45–46.
95. Carter, *Breakthrough*, 100.
96. J. F. Enders and W. M. Hammon, "Active and Passive Immunization against the Virus of Malignant Pan Leucopenia of Cats," *Proceedings of the Society for Experimental Biology and Medicine* 48 (1940): 194–200.
97. "Background Story," 1952, S 3: GG FT, Box 3, SSR, MDA, 10; Fowler, "Dr. William Hammon Dies at 85." See also *Preventive Medicine in World War II*, vol. 5, *Communicable Diseases Transmitted through Contact or by Unknown Means, U.S. Army Medical Department* (Falls Church, Va.: Office of the Surgeon General, 1960), chapter 15, 382, http://history.amedd.army.mil/booksdocs/wwii/communicablediseasesV5/chapter15.htm, last viewed May 23, 2014; Leonard D. Heaton, W. Paul Havens, John Boyd Coates Jr., *Internal Medicine in World War II*, vol. 2: *Infectious Diseases, Medical Department, United States Army* (Washington, D.C.: Office of the Surgeon General Department of the Army, 1963), 82; Woodward, *The Armed Forces Epidemiological Board*, 15–16.
98. Reeves, interview, 45–46; Carter, *Breakthrough*, 100.
99. Patricia M. Hammer, *The Decade of Elusive Promise: Professional Women in the United States, 1920–1930* (Ann Arbor, Mich.: University Microfilms International, 1979); Regina Morantz-Sanchez, *Sympathy and Science: Women Physicians in American Medicine* (New York: Oxford University Press, 1985); Margaret Rossiter, *Women Scientists in America: Struggles and Strategies to 1940* (Baltimore: Johns Hopkins University Press, 1982); Harriet Zuckerman, "Stratification in American Science," *Sociological Inquiry* 40 (Spring 1979): 235–257.
100. Reeves, interview, 45–46.
101. William Hammon and Beatrice Howitt, "Epidemiological Aspects of Encephalitis in the Yakima Valley, Washington: Mixed St. Louis and Western Equine Types," *American Journal of Hygiene* 35, 2 (March 1942): 163–185.
102. W. McD. Hammon and W. C. Reeves, "Laboratory Transmission of St. Louis Encephalitis Virus by Three Genera of Mosquitoes," *Journal of Experimental Medicine* 78, 4 (October 1943): 241–253; W. McD. Hammon and W. C. Reeves, "Laboratory Transmission of Western Equine Encephalomyelitis Virus by Mosquitoes of the Genera *Culex* and *Culiseta*," *Journal of Experimental Medicine* 78, 6 (December 1943): 425–434; D. M. Goldstein, W. McD. Hammon, and H. R. Viets, "An Outbreak of Polioencephalitis among Navy Cadets, Possibly Food Borne," *JAMA* 131 (June 1946): 569–573; W. McD. Hammon and W. C. Reeves, "Interepidemic Studies on Arthropod-Borne Virus Encephalitides and Poliomyelitis in Kern County, California, and the Yakima Valley, Washington, 1944," *American Journal of Epidemiology* 46, 3 (1947): 326–335.
103. Reeves, interview, 66–67.
104. "News From the Field: Faculty Appointments, School of Public Health, University of California," *AJPH* 35 (October 1945): 1108.

105. Hammon to Francis, February 27, 1945, "H," Box 4, Thomas Francis Papers, BHL.
106. Hammon to Francis, April 10, 1945, ibid.
107. "Background Story," 1952, S 3: GG FT, Box 3, SSR, MDA, 10; Fowler, "Dr. William Hammon Dies at 85."
108. Woodward, *The Armed Forces Epidemiological Board*, 87.
109. "Office of the Surgeon General," Office of Public Health and Science, U.S. Department of Health and Human Services, http://www.surgeongeneral.gov/about/previous/bioparran.htm, last viewed October 2010; Rinaldo, "Passive Immunization," 790.
110. Rinaldo, "Passive Immunization," 791.
111. "News and Notes: San Francisco," *California Medicine* 71, 5 (November 1949): 377; "Biographical Sketches of Dr. William McD. Hammon and Assistants," 1951, S 3: GG FT, Box 3, SSR, MDA, 6–7.
112. Rinaldo, "Passive Immunization," 791.
113. Reeves, interview, 57.
114. "Polio Expert Says Peak of Epidemic Is Past in Sioux Falls," *The Huronite and the Daily Plainsman*, October 21, 1948, 1.
115. John R. Paul, *A History of Poliomyelitis* (New Haven: Yale University Press, 1971), 391–392.
116. Rinaldo, "Passive Immunization," 791.
117. Dr. William McD. Hammon, "Address to the Medical Society," August 1951, S 3: GG FT, Box 3, SSR, MDA.
118. Hammon to Sabin, November 2, 1949, Hammon, W. McD 1944–64, Folder 22, Box 11, Hauck Center for the Albert B. Sabin Archives, University of Cincinnati Libraries, Cincinnati, Ohio (henceforth HCASA).
119. Sabin to Hammon, November 4, 1949, Hammon, W. McD 1944–64, Folder 22, Box 11, HCASA.
120. Hammon, "Address to the Medical Society," August 1951, S 3: GG FT, Box 3, SSR, MDA.
121. John F. Enders, Thomas H. Weller, and Frederick C. Robbins, "Cultivation of the Lansing Strain of Poliomyelitis Virus in Cultures of Various Human Embryonic Tissues," *Science* 109, 2822 (January 1949): 85–87; Williams, "Chapter VIII: Passive Immunity," Preliminary Draft, 1956, S 1: Passive Immunity, Box 1, HNR, MDA, 13; Oshinsky, *Polio: An American Story*, 123–124.
122. D. Bodian, "Neutralization of Three Immunological Types of Poliomyelitis Virus by Human Gamma Globulin," *Proceedings of the Society for Experimental Biology and Medicine* 72 (1949): 259–261.
123. Rinaldo, "Passive Immunization," 791.
124. William McD. Hammon, "Possibilities of Specific Prevention and Treatment of Poliomyelitis," *Pediatrics* 6, 5 (November 1950): 696–705.
125. Hammon, "Address to the Medical Society," August 1951, S 3: GG FT, Box 3, SSR, MDA.
126. J. R. Paul and J. T. Riordan, "Observations on Serological Epidemiology: Antibodies to the Lansing Strain of Poliomyelitis Virus in Sera from Alaskan Eskimos," *American Journal of Epidemiology* 52 (September 1950): 202–212.
127. William McD. Hammon, "Immunity in Poliomyelitis," *Bacteriological Reviews* 13, 3 (September 1949): 136, 138. For more on waning immunity, see Philip Cohen, Herman Schneck, Emanuel Dubow, Sidney Q. Cohlan, "The Changed Status of Diphtheria Immunity," *Pediatrics* 3, 5 (May 1949): 630–638.

128. W. M. Hammon, E. H. Ludwig, G. E. Sather, and W. D. Schrack Jr., "A Longitudinal Study of Infection with Poliomyelitis Viruses in American Families on a Philippine Military Base during an Inter-Epidemic Period," *Annals of the New York Academy of Sciences* 61 (September 1955): 979–988.
129. For more about the 1899 documented outbreak on Guam, see Mark A. Nordenberg, *Defeat of an Enemy: Chancellor Mark A. Nordenberg Reports on the 50th Anniversary Celebration of the Triumph of the Pitt Polio Vaccine* (Pittsburgh: University of Pittsburgh, 2005), 40–41.
130. Hammon, "Immunity in Poliomyelitis," 137.
131. Ibid., 139, 148.
132. Ibid., 154.
133. Robert F. Rogers, *Destiny's Landfall: A History of Guam* (Honolulu: University of Hawai'i Press, 1995); David Arnold, *Colonizing the Body: State Medicine and Epidemic Disease in Nineteenth-Century India* (Berkeley: University of California Press, 1993); Laura Briggs, *Reproducing Empire: Race, Sex, Science, and US Imperialism in Puerto Rico* (Berkeley: University of California Press, 2002).
134. Hammon, "Immunity in Poliomyelitis," 143.
135. "Meeting on Gamma Globulin," February 3, 1950, S 3: Conferences, Box 5, CMR, MDA, 78.
136. Ibid., 77–78.
137. Rinaldo, "Passive Immunization," 791–792; "Meeting on Gamma Globulin," February 3, 1950, S 3: Conferences, Box 5, CMR, MDA, 77–83; Isabel Morgan, "Immunization of Monkeys with Formalin-Inactivated Poliomyelitis Virus," *American Journal of Hygiene* 48 (1948): 394–410.
138. "Meeting on Gamma Globulin," February 3, 1950, S 3: Conferences, Box 5, CMR, MDA, 85–86; Byron Spice, "Tireless Polio Research Effort Bears Fruit and Indignation, The Salk Vaccine: 50 Years Later," *Pittsburgh Post-Gazette*, April 4, 2005; Rinaldo, "Passive Immunization," 791.
139. Amy L. Fairchild and Gerald M. Oppenheimer, "Public Health Nihilism vs. Pragmatism: History, Politics, and the Control of Tuberculosis," *AJPH* 88, 7 (July 1998): 1105–1117.
140. Rinaldo, "Passive Immunization," 791.
141. William McD. Hammon, "Suggested Plans for a Field Trial to Determine the Effectiveness of Gamma Globulin in the Prophylaxis of Poliomyelitis," April 1951, Folder 4, Box 244, Dr. Jonas E. Salk Papers (1926–1991), Mandeville Special Collections Library, La Jolla, California (henceforth MSCL), 10; Susan M. Reverby, "'Normal Exposure' and Inoculation Syphilis: A PHS 'Tuskegee' Doctor in Guatemala, 1946–48," *Journal of Policy History* 23, 1 (January 2011): 6–28.
142. Reverby, "'Normal Exposure' and Inoculation Syphilis."
143. Hammon, "Immunity in Poliomyelitis," 143; "Proceedings of the Committee on Immunization of the National Foundation for Infantile Paralysis," May 17, 1951, Folder 4, Box 244, Dr. Jonas E. Salk Papers (1926–1991), MSCL.
144. Harry Marks, *The Progress of Experiment: Science and Therapeutic Reform in the United States, 1900–1990* (Cambridge: Cambridge University Press, 1997, 2000), 141–146; Trevor Pinch, "'Testing—One, Two, Three . . . Testing!': Toward a Sociology of Testing," *Science, Technology, & Human Values* 18 (Winter 1993): 29–30; V. Farewell and T. Johnson, "Woods and Russell, Hill, and the Emergence of Medical Statistics," *Statistics in Medicine* 29, 14 (June 2010): 1459–1476.

145. William McD. Hammon, Lewis L. Coriell, and Joseph Stokes Jr., "Evaluation of Red Cross Gamma Globulin as a Prophylactic Agent for Poliomyelitis: 1," *JAMA* 150, 8 (October 25, 1952): 742.
146. Hammon, "Suggested Plans for a Field Trial," April 1951, Folder 4, Box 244, Dr. Jonas E. Salk Papers (1926–1991), MSCL, 13.

Chapter 2 – Building Consent for a Clinical Trial

1. "Meeting on Gamma Globulin, Waldorf Astoria, New York City," February 3, 1950, S 3: Conferences, Box 5, Conferences and Meetings Records (henceforth CMR), March of Dimes Archives (henceforth MDA), White Plains, New York, 157.
2. Walter Lippmann, *Public Opinion* (New York: Macmillan, 1922).
3. Charles R. Rinaldo, "Passive Immunization against Poliomyelitis: The Hammon Gamma Globulin Field Trials, 1951–1953," *AJPH* 95, 5 (May 2005), 793.
4. Robert Bud, "Germophobia to the Carefree Life," in *Medicating Modern America: Prescription Drugs in History*, ed. Andrea Tone and Elizabeth Siegel Watkins (New York: New York University Press, 2007), 17–41; "Meeting on Gamma Globulin," February 3, 1950, S 3: Conferences, Box 5, CMR, MDA, 85–86.
5. "News and Notes: San Francisco," *California Medicine* 71, 5 (November 1949): 377; "Biographical Sketches of Dr. William McD. Hammon and Assistants," 1951, S 3: GG FT, Box 3, Surveys and Studies Records (henceforth SSR), MDA, 6–7. See also Rinaldo, "Passive Immunization," 791.
6. "News and Notes: San Francisco," 377; Rinaldo, "Passive Immunization," 791.
7. Richard Carter, *Breakthrough: The Saga of Jonas Salk* (New York: Pocket Books, 1967), 99–100.
8. Jane Smith, *Patenting the Sun: Polio and the Salk Vaccine* (New York: William Morrow, 1990), 143; Carter, *Breakthrough*, 100.
9. Dr. Julius S. Youngner, e-mail message to author, May 13, 2010.
10. Byron Spice, "Tireless Polio Research Effort Bears Fruit and Indignation," *Pittsburgh Post-Gazette*, April 4, 2005, http://www.post-gazette.com/pg/05094/482468.stm, last viewed October 1, 2010.
11. Carter, *Breakthrough*, 101.
12. Dr. Julius S. Youngner, e-mail message to author, May 13, 2010.
13. Joseph Stokes Jr., Irving J. Wolman, Howard Childs Carpenter, Julius Margolis, "Prophylactic Use Of Parents' Whole Blood in Anterior Poliomyelitis: Philadelphia Epidemic of 1932," *American Journal of Diseases of Children* 50, 3 (1935): 581–595; J. F. Kessel, A. S. Hoyt, and R. T. Fisk, "Use of Serum and the Routine and Experimental Laboratory Findings in the 1934 Poliomyelitis Epidemic," *AJPH* 24 (1934): 1215–1223; J. Stokes Jr., E. P. Maris, and S. S. Gelliss, "Chemical, Clinical, and Immunological Studies on the Products of Human Plasma Fractionation. XI. Use of Concentrated Normal Human Serum Gamma Globulin (Human Immune Serum Globulin) in the Prophylaxis and Treatment of Measles," *Journal of Clinical Investigation* 23 (1944): 531–540. Stokes to Philadelphia Hospital Board of Managers, April 24, 1952, St65p, NFIP #6, Joseph Stokes, Jr., Papers, American Philosophical Society, Philadelphia, Pennsylvania (henceforth APS).
14. Stokes to Hammon, February 10, 1950, St65p, Hammon, W. McD. #1, Joseph Stokes, Jr., Papers, APS.
15. John R. Paul, *A History of Poliomyelitis* (New Haven: Yale University Press, 1971), 392.
16. William McD. Hammon, "Possibilities of Specific Prevention and Treatment of Poliomyelitis," *Pediatrics* 6, 5 (November 1950): 696–705.

17. Murphy to Stokes, September 1, 1950, St65p, Poliomyelitis #20, Joseph Stokes, Jr., Papers, APS.
18. Stokes to Murphy, September 15, 1950, ibid.
19. "Meeting on Gamma Globulin," February 3, 1950, S 3: Conferences, Box 5, CMR, MDA, 141.
20. Kenton Kroker, Jennifer Keelan, and Pauline M. H. Mazumdar, eds., *Crafting Immunity: Working Histories of Clinical Immunology* (Aldershot, UK: Ashgate Publishing, 2008).
21. Hilary Koprowski, "First Decade (1950–1960) of Studies and Trials with the Polio Vaccine," *Biologicals* 34 (2006): 82.
22. Hilary Koprowski, George A Jervis, Thomas W. Norton, "Immune Responses in Human Volunteers upon Oral Administration of a Rodent Adapted Strain of Poliomyelitis Virus," *American Journal of Hygiene* 55 (1952): 108–126; Edward Hooper, *The River: A Journey to the Source of HIV and AIDS* (Boston: Little, Brown, 1999).
23. Koprowski, "First Decade (1950–1960)," 82.
24. David M. Oshinsky, *Polio: An American Story: The Crusade That Mobilized the Nation against the 20th Century's Most Feared Disease* (Oxford: Oxford University Press, 2005), 159.
25. William McD. Hammon, "Suggested Plans for a Field Trial to Determine the Effectiveness of Gamma Globulin in the Prophylaxis of Poliomyelitis," April 1951, Box 244, Folder 4, Dr. Jonas E. Salk Papers (1926–1991), Mandeville Special Collections Library, La Jolla, California (henceforth MSCL).
26. "Thousands Enroll in Paralysis Drive," *New York Times*, January 20, 1938, 17; "President Urges Paralysis Gifts," *New York Times*, November 24, 1941, 13; Oshinsky, *Polio: An American Story*, 65.
27. Stephen E. Mawdsley, "Harnessing the Power of People: The Fundraising Efforts of the National Foundation for Infantile Paralysis, 1938–1945" (BA honors thesis, University of Alberta, 2006).
28. "Progress Made in Polio Battle," *Statesville (North Carolina) Daily Record*, June 14, 1949, 4; "News Story," September 5, 1951, S 14: Polio, Box 12, Medical Program Records (henceforth MPR), MDA; Editorial, "An Object Worth All the Dimes in Sight," *San Antonio Express*, December 14, 1950, 6.
29. "Radio News," *Oakland Tribune*, June 27, 1948, 22.
30. Foster Rhea Dulles, *The American Red Cross* (New York: Harper and Brothers, 1950); Patrick F. Gilbo, *The American Red Cross: The First Century* (New York: Harper & Row, 1981); Julia Irwin, *Making the World Safe: The American Red Cross and a Nation's Humanitarian Awakening* (New York: Oxford University Press, 2013).
31. G. Foard McGinnes, Ross T. McIntire, and George W. Hervey, "The National Blood Program of the American Red Cross," *AJPH* 39, 11 (November 1949): 1429–1433.
32. "Meeting on Gamma Globulin," February 3, 1950, S 3: Conferences, Box 5, CMR, MDA, 182–183.
33. Williams, "Chapter VIII: Passive Immunity to Poliomyelitis," Preliminary Draft, 1956, S 1: Passive Immunity, Box 1, History of the NFIP Records, MDA, 29.
34. For more about the purpose and structure of the Committee on Immunization, see "Proceedings of the Committee on Immunization," May 17, 1951, Folder 4, Box 244, Dr. Jonas E. Salk Papers (1926–1991), MSCL, 1.
35. "Meeting on Gamma Globulin," February 3, 1950, S 3: Conferences, Box 5, CMR, MDA, 157.

36. Ibid., 105.
37. Ibid., 93, 110–111.
38. "Proceedings of the Committee," May 17, 1951, Folder 4, Box 244, Dr. Jonas E. Salk Papers (1926–1991), MSCL, 8–9, 11.
39. Ibid.,18.
40. Ibid., 22.
41. Rivers in Saul Benison, *Tom Rivers: Reflections on a Life in Medicine and Science* (Cambridge, Mass.: MIT Press, 1967), 479.
42. Andrea Rusnock, "Making Sense of Vaccination c. 1800," in Kroker, Keelan, and Mazumdar, *Crafting Immunity* (Aldershot, UK: Ashgate Publishing, Ltd., 2008); Mark Jackson, "'A Private Line to Medicine': The Clinical and Laboratory Contours of Allergy in the Early Twentieth Century" in *Crafting Immunity.*
43. "Proceedings of the Committee," May 17, 1951, Folder 4, Box 244, Dr. Jonas E. Salk Papers (1926–1991), MSCL, 4.
44. Ibid., 9; "Proceedings of the Committee," July 6, 1951, ibid., 29.
45. "Proceedings of the Committee," May 17, 1951, ibid., 7, 26.
46. Ibid., 19.
47. Rivers, in Benison, *Tom Rivers*, 480.
48. Elizabeth W. Etheridge, *Sentinel for Health: A History of the Centers for Disease Control* (Berkeley: University of California Press, 1992); Mark E. Rushefsky and Deborah R. McFarlane, *The Politics of Public Health in the United States* (Armonk, N.Y.: M. E. Sharpe, 2005).
49. "Specter of Paralysis Stalks Carolina," *Literary Digest*, July 1935; Heather Green Wooten, *The Polio Years in Texas: Battling a Terrifying Unknown* (College Station: Texas A&M University Press, 2009), 91; Naomi Rogers, *Dirt and Disease: Polio before FDR* (New Brunswick, N.J.: Rutgers University Press, 1992); Smith, *Patenting the Sun*, 35–36.
50. "Frame Rules to Check Infantile Paralysis," *New York Times*, July 29, 1944, 10; "Rules to Prevent Paralysis Given," *New York Times*, June 1, 1949, 41; "Five Polio Precautions Are Listed for Parents," *Emmetsburg (Iowa): Reporter*, April 12, 1949, 2.
51. "First Polio Reports Emphasize Need for Observing Precautions," *Robesonian* (Lumberton, North Carolina), July 12, 1951, 4.
52. David Rose, *Images of America: March of Dimes* (Charleston, S.C.: Arcadia Publishing, 2003), 34; "1951 Polio Pointers," *Popular Science*, September 1951, 60.
53. "Polio Precautions," *Independent Journal* (California), July 12, 1951, 18.
54. "Polio Pointers," *Chillicothe Constitution-Tribune* (Missouri), September 6, 1951, 3.
55. Becker to Weaver, October 11, 1951, S 14: Polio, Box 12, MPR, MDA.
56. "Proceedings of the Committee," May 17, 1951, Folder 4, Box 244, Dr. Jonas E. Salk Papers (1926–1991), MSCL, 37.
57. Ibid., 41.
58. Ibid., 37–41.
59. Ibid., 54.
60. Ibid., 42.
61. Stephen E. Mawdsley, "Balancing Risks: Childhood Inoculations and America's Response to the Provocation of Paralytic Polio," *Social History of Medicine* 26, 4 (November 2013): 759–778. Researchers later discovered the mechanism behind polio provocation; see Matthias Gromeier and Eckard Wimmer, "Mechanism of Injury-Provoked Poliomyelitis," *Journal of Virology* 72, 6 (June 1998): 5056–5060.

62. Biographical Sketch of Gaylord W. Anderson (1901–1979), Gaylord W. Anderson papers, University Archives, University of Minnesota, Twin Cities.
63. Gaylord W. Anderson and Audrey E. Skaar, "Poliomyelitis Occurring after Antigen Injections," *Pediatrics* 7, 6 (June 1951): 741–759.
64. Editorial, *Journal of Pediatrics* 38, 6 (June 1951): 781–782.
65. Harold K. Faber, "Postinoculation Poliomyelitis," *Pediatrics* 7, 2 (1951): 300–304.
66. David M. Herszenhorn, "Robert F. Korns, 82, Researcher Who Helped Test Polio Vaccine," *New York Times*, October 16, 1995.
67. Harry Weaver, Editorial, June 1951, S 3: GG FT, Box 3, SSR, MDA; Robert F. Korns, Robert M. Albrecht, and Frances B. Locke, "The Association of Parenteral Injections with Poliomyelitis," *AJPH* 42, 2 (February 1952): 153–169; Robert M. Albrecht and Frances B. Locke, "Effect of Physical Activity on Prognosis of Poliomyelitis," *JAMA* 146, 9 (1951): 769–771.
68. "Proceedings of the Committee," July 6, 1951, Box 244, Folder 4, Dr. Jonas E. Salk Papers, MSCL, 1.
69. Ibid., 17.
70. Jens I. Zinn, "Risk as Discourse: Interdisciplinary Perspectives," *Critical Approaches to Discourse Analysis across Disciplines* 4, 2 (2010): 106–124.
71. "Proceedings of the Committee," May 17, 1951, Folder 4, Box 244, Dr. Jonas E. Salk Papers (1926–1991), MSCL, 51.
72. "U.S. Decries 'Shots' in 'Polio Season,'" *New York Times*, June 22, 1951, 27.
73. Hilleboe to State Public Health Physicians, June 11, 1951, S 3: GG FT, Box 3, SSR, MDA; "State Acts to Curb Polio Spread Peril," *New York Times*, June 13, 1951, 30.
74. Rivers, in Benison, *Tom Rivers*, 477.
75. "Proceedings of the Committee," May 17, 1951, Folder 4, Box 244, Dr. Jonas E. Salk Papers (1926–1991), MSCL, 8, 12.
76. Ibid., 13, 52, 56.
77. Ibid., 8.
78. Rivers, in Benison, *Tom Rivers*, 478.
79. "Proceedings of the Committee," May 17, 1951, Folder 4, Box 244, Dr. Jonas E. Salk Papers (1926–1991), MSCL, 8, 13.
80. Paul, *A History of Poliomyelitis*, 393.
81. "Proceedings of the Committee," May 17, 1951, Folder 4, Box 244, Dr. Jonas E. Salk Papers (1926–1991), MSCL, 62.
82. Douglas Starr, *Blood* (New York: Alfred Knopf, 1998), 212–213.
83. Jules L. Dienstag, "Chapter 298. Acute Viral Hepatitis," in *Harrison's Principles of Internal Medicine*, 17th ed., ed. Anthony S. Fauci, Eugene Braunwald, Dennis L. Kasper, Stephen L. Hauser, Dan L. Longo, J. Larry Jameson, and Joseph Loscalzo (New York: McGraw-Hill Medical Publishing Division, 2008).
84. Jacalyn Duffin, *Lovers and Livers: Disease Concepts in History* (Toronto: University of Toronto Press, 2005), 85.
85. Lawrence S. Friedman, "Chapter 16. Liver, Biliary Tract and Pancreas Disorders," in *Current Medical Diagnosis and Treatment 2009*, ed. Stephen J. McPhee and Maxine A. Papadakis (New York: McGraw-Hill Medical Publishing Division, 2008).
86. Jenny Stanton, "'I've Been on Tenterhooks': Wartime Medical Research Council Jaundice Committee Experiments," in *Useful Bodies: Humans in the Service of Medical Science in the Twentieth Century*, ed. Jordan Goodman, Anthony McElligott, and Lara Marks (Baltimore: Johns Hopkins University Press, 2003), 110, 121.

87. Ross L. Gauld, "Field Studies Relating to Immunity in Infectious Hepatitis and Homologous Serum Jaundice," *AJPH* 37 (April 1947): 400–406; Duffin, *Lovers and Livers*, 90.
88. James W. Colbert, "Review of Animal Experimentation in Infectious Hepatitis and Serum Hepatitis," *Yale Journal of Biology and Medicine* 21, 4 (March 1949): 335–343.
89. Charles L. Hoagland, "An Analysis of the Effect of Fat in the Diet on Recovery in Infectious Hepatitis," *AJPH* 36 (November 1946): 1287–1292; O. Gertzen, "Diet in the Treatment of Acute Hepatitis," *British Medical Journal* 20, 1 (May 1950): 1166–1168; Thomas C. Chalmers, Richard D. Eckhardt, William E. Reynolds, Joaquin G. Cigarroa, Jr., Norman Deane, Robert W. Reifenstein, Clifford W. Smith, Charles S. Davidson, Mary A. Maloney, Merme Bonnel, Marian Niiya, Alice Stang, and Ann McD. O'Brien, "The Treatment of Acute Infectious Hepatitis. Controlled Studies of the Effects of Diet, Rest, and Physical Reconditioning on the Acute Course of the Disease and on the Incidence of Relapses and Residual Abnormalities," *Journal of Clinical Investigation* 34 (July 1955): 1163–1235.
90. Henry J. Nichols, "Agglutination of Typhoid Group of Organisms in Cases of Jaundice among Vaccinated Persons," *JAMA* 81, 23 (December 1923): 1946–1948; G. M. Findlay and F. O. MacCallum, "Hepatitis and Jaundice Associated with Immunization against Certain Virus Diseases: (Section of Comparative Medicine)," *Proceedings of the Royal Society of Medicine* 31, 7 (May 1938): 799–806; British Ministry of Health, "Homologous Serum Jaundice: A Memorandum," *Lancet* 1, 16 (January 1943): 83–88; "Discussion on Infective Hepatitis, Homologous Serum Hepatitis, and Arsenotherapy Jaundice," *Proceedings of the Royal Society of Medicine* 37, 8 (June 1944): 449–460; John R. Neefe, Joseph Stokes Jr., John G. Reinhold, and F.D.W. Lukens, "Hepatitis Due to the Injection of Homologous Blood Products in Human Volunteers," *Journal of Clinical Investigation* 23, 5 (September 1944): 836–855.
91. John Farley, *To Cast Out Disease: A History of the International Health Division of the Rockefeller Foundation, 1913–1951* (New York: Oxford University Press, 2004), 170; Wilbur A. Sawyer to Margaret Sawyer, March 24, 1942, Folder 17, Box 2, Wilbur A. Sawyer Papers, National Library of Medicine, Bethesda, Maryland; "New Vaccine Boom to Army in Jungle," *New York Times*, April 7, 1943, 27.
92. Farley, *To Cast Out Disease*, 173.
93. *Staff Study Report on the National Blood Program*, January 1952, American Red Cross Archives, Washington, D.C., 32; Charles A. Janeway, "Blood and Blood Derivatives—A New Public Health Field," *AJPH* 36, 1 (January 1946), 3; Waldemar Kaempffert, "Science in Review: Cause of Army Jaundice Is Now Discovered and the Means of Control Indicated," *New York Times*, January 21, 1945, 71.
94. W. Chas. Cockburn, John A. Harrington, Reginald A. Zeitlin, David Morris, and Francis E. Camps, "Homologous Serum Hepatitis and Measles Prophylaxis: A Report to the Medical Research Council," *British Medical Journal*, July 7, 1951, 6–12.
95. Cockburn et al., "Homologous Serum Hepatitis and Measles Prophylaxis," 6–12.
96. George T. Pollock, *Fevers and Cultures: Lessons for Surveillance, Prevention and Control* (Abingdon, UK: Radcliffe Medical Press, 2003), 26.
97. "Henry W. Kumm, 89, Foundation Official," *New York Times*, January 15, 1991, D 19; Kumm, Henry W., Biographical Data, Series: Human Resources, Box 9, MPR, MDA; Henry W. Kumm and Thomas B. Turner, "The Transmission of Yaws from Man to Rabbits by an Insect Vector, Hippelates Pallipes Loew," *American Journal*

of Tropical Medicine and Hygiene (1936): 245–271; National Foundation News Release, September 2, 1953, Kumm, Henry W., Biographical Data, Series: Human Resources, Box 9, MPR, MDA.

98. Kumm to Weaver, 1951, S 3: GG FT, Box 3, SSR, MDA.
99. U.S. Army Medical Research and Materiel Command (USAMRMC), "Joseph E. Smadel," http://wrair-www.army.mil/images/Dr-Joseph-E-Smadel.pdf, last viewed August 18, 2009.
100. Smadel to Hammon, July 26, 1951, S 3: GG FT, Box 3, SSR, MDA.
101. L. C. Kolb and S. J. Gray, "Peripheral Neuritis as a Complication of Penicillin Therapy," *JAMA* 132 (1946): 323–326; N. Howard-Jones, "A Critical Study of the Origins and Development of Hypodermic Medication," *Journal of the History of Medicine and Allied Sciences* 2 (1947): 201–249; T. R. Broadbent, G. L. Odom, and B. Woodall, "Peripheral Nerve Injuries from Administration of Penicillin," *JAMA* 140 (1949): 1008–1010; J. Knowles, "Accidental Intra-Arterial Injection of Penicillin," *American Journal of Diseases of Children* 111 (1966): 552–556; K. Clark, P. Williams, and W. Willis, "Injection Injury of the Sciatic Nerve," *Clinical Neurosurgery* 17 (1969): 111–124; F. H. Gilles and D. D. Matson, "Sciatic Nerve Injury Following Misplaced Gluteal Injection," *Journal of Pediatrics* 76, 2 (February 1970): 247–254; O. Svendsen, "Intramuscular Injections and Muscle Damage: Effects of Concentration, Volume, Injection Speed, and Vehicle," *Archives of Toxicology* 7 (1984): 472–475.
102. F. P. Hudson, Anne McCandless, and A. G. O'Malley, "Sciatic Paralysis in Newborn Infants," *British Medical Journal* 1, 4647 (1950): 223–225.
103. "Proceedings of the Committee," May 17, 1951, Folder 4, Box 244, Dr. Jonas E. Salk Papers (1926–1991), MSCL, 25.
104. M. A. Combs, W. K. Clark, C. F. Gregory, and J. A. James, "Sciatic Nerve Injury in Infants," *JAMA* 173 (1960): 1336; Jeanette M. Daly, William Johnston, and Younghae Chung, "Injection Sites Utilized for DPT Immunizations in Infants," *Journal of Community Health Nursing* 9, 2 (1992): 88.
105. Allan M. Brandt, "Polio, Politics, Publicity, and Duplicity: Ethical Aspects in the Development of the Salk Vaccine," *International Journal of Health Services* 8, 2 (1978): 257–270.
106. Lippmann, *Public Opinion*, 158.
107. "Proceedings of the Committee," July 6, 1951, Folder 4, Box 244, Dr. Jonas E. Salk Papers (1926–1991), MSCL.
108. Ibid., 17.
109. Ibid., 18–19, 21.
110. Ibid., 25.
111. Ibid., 21.
112. Naomi Oreskes and Erik M. Conway, *Merchants of Doubt: How a Handful of Scientists Obscured the Truth on Issues from Tobacco Smoke to Global Warming* (New York: Bloomsbury, 2010).
113. "Proceedings of the Committee," May 17, 1951, Folder 4, Box 244, Dr. Jonas E. Salk Papers (1926–1991), MSCL, 20.
114. Ibid., 32.
115. "Proceedings of the Committee," July 6, 1951, ibid., 24.
116. Ibid., 28.
117. Susan E. Lederer, *Subjected to Science: Human Experimentation in America before the Second World War* (Baltimore: Johns Hopkins University Press, 1997), Introduction.

118. "Proceedings of the Committee," May 17, 1951, Folder 4, Box 244, Dr. Jonas E. Salk Papers (1926–1991), MSCL, 52.
119. Stanley Milgram, *Obedience to Authority: An Experimental View* (New York: Harper & Row, 1974).
120. Smith, *Patenting the Sun*, 63.
121. "Proceedings of the Committee," July 6, 1951, Folder 4, Box 244, Dr. Jonas E. Salk Papers (1926–1991), MSCL, 30.
122. Solomon E. Asch, "Opinions and Social Pressure," *Scientific American* 193 (1955): 31–35; Solomon E. Asch, "Effects of Group Pressure Upon the Modification and Distortion of Judgments," in *Documents of Gestalt Psychology*, ed. Mary Henle (Berkeley: University of California Press, 1961).
123. "Proceedings of the Committee," July 6, 1951, Folder 4, Box 244, Dr. Jonas E. Salk Papers (1926–1991), MSCL, 31.
124. Paul, *A History of Poliomyelitis*, 393.
125. Stokes to Hammon, August 13, 1951, St65p, Hammon, W. McD. #5, Joseph Stokes, Jr., Papers, APS.
126. Hammon to Stokes, August 16, 1951, ibid.
127. Weaver to Hilleboe, August 21, 1951, S 3: GG FT, Box 3, SSR, MDA.
128. W. M. Hammon, L. L. Coriell, and J. Stokes Jr., "Evaluation of Red Cross Gamma Globulin as a Prophylactic Agent for Poliomyelitis: 1. Plan of Controlled Field Tests and Results of the 1951 Pilot Study in Utah," *JAMA* 150 (1952): 742; Kumm to Hammon, August 10, 1951, S 3: GG FT, Box 3, SSR, MDA.
129. Hammon et al., "Evaluation of Red Cross Gamma Globulin," Part 1, 743.
130. H. K. Faber, "Postinoculation Poliomyelitis," *Pediatrics* 7, 2 (February 1951): 300–304.
131. Hammon et al., "Evaluation of Red Cross Gamma Globulin," Part 1, 742.
132. "Worker's Badly Needed," *Houston Informer*, July 5, 1952, 1.
133. Hammon et al., "Evaluation of Red Cross Gamma Globulin," Part 1, 743.
134. Ibid.
135. "Report of Gamma Globulin Passive Immunization Field Project, Utah County, Utah, 1951," November 30, 1951, S 14: Polio, Box 12, MPR, MDA, 11.
136. Hammon to Kumm, December 27, 1951, S 3: GG FT, Box 3, SSR, MDA; "Report of Gamma Globulin," November 30, 1951, S 14: Polio, Box 12, MPR, MDA, 11; Hammon, Coriell, Cheever, Stokes, April 2, 1952, "Cumulated Incidence of Reported Disease by Week, Month, or Twenty-Week Period after Injection of Gamma Globulin or Gelatin—Final Report," S 3: GG FT, Box 3, SSR, MDA; Hammon to NFIP, March 5, 1952, S 3: GG FT, Box 3, SSR, MDA.
137. Hammon, "Draft Reply Card," January 7, 1952, S 3: GG FT, Box 3, SSR, MDA.
138. Hammon to Kumm, December 27, 1951, S 3: ibid.
139. Hammon, "Draft Reply Card," ibid.
140. Hammon to Kumm, December 27, 1951, S 3: ibid.
141. Kumm to Hammon, December 31, 1951, S 3: ibid.

Chapter 3 – Marketing and Mobilization

1. Van Riper to O'Connor, June 21, 1951, S 14: Polio, Box 12, Medical Program Records (henceforth MPR), March of Dimes Archives (henceforth MDA).
2. Johannes Ipsen, "Administrative Problems in the Use of Poliomyelitis Immune Globulin," *AJPH* 43 (September 1953): 1101–1110.
3. Van Riper to O'Connor, June 21, 1951, S 14: Polio, Box 12, MPR, MDA.

4. "Biographical Sketches of Dr. William McD. Hammon and Assistants," 1951, S 3: GG FT, Box 3, Surveys and Studies Records (henceforth SSR), MDA, 7. Nancy Hulston, *One Hundred Firsts: University of Kansas School of Medicine* (Lawrence: University of Kansas, 2005), 4.
5. "Cheever, Francis Sargent, M.D.," *New York Times*, September 23, 1997, Obituaries; "Background Story, Dr. F. S. Cheever," S 3: GG FT, Box 3, SSR, MDA, 8.
6. W. M. Hammon, L. L. Coriell, and J. Stokes Jr., "Evaluation of Red Cross Gamma Globulin as a Prophylactic Agent for Poliomyelitis: 1. Plan of Controlled Field Tests and Results of the 1951 Pilot Study in Utah," *JAMA* 150 (1952): 742; Kumm to Hammon, August 10, 1951, S 3: GG FT, Box 3, SSR, MDA.
7. "Proceedings of the Committee on Immunization of the National Foundation for Infantile Paralysis," July 6, 1951, Box 244, Folder 4, Dr. Jonas E. Salk Papers, Mandeville Special Collections Library (henceforth MSCL), California, 30.
8. Thomas J. Sugrue, *The Origins of the Urban Crisis: Race and Inequality in Postwar Detroit* (Princeton: Princeton University Press, 2005); Judith Goode and Jo Anne Schneider, *Reshaping Ethnic and Racial Relations in Philadelphia* (Philadelphia: Temple University Press, 1994).
9. "Proceedings of the Committee," July 6, 1951, Box 244, Folder 4, Dr. Jonas E. Salk Papers, MSCL, 30.
10. Arthur J. Vidich and Joseph Bensman, *Small Town in Mass Society: Class, Power, and Religion in a Rural Community* (Princeton: Princeton University Press, 1968), chapter 2; Granville Hicks, *Small Town* (New York: Fordham University Press, 2004), chapter 8; Glenn V. Fuguitt, David L. Brown, and Calvin L. Beale, *Rural and Small Town America* (New York: Russell Sage Foundation, 1989), chapter 12.
11. "Country Doctor," *Life* magazine, September 20, 1948, 115–126; Sasha Mullally, *Unpacking the Black Bag: A History of North American Country Doctors, 1900–1950* (Toronto: University of Toronto Press, 2009).
12. "Infantile Paralysis: Child Victims Fill Beds of an Emergency Hospital as Epidemic Hits Rural Counties of North Carolina," *Life* magazine, July 21, 1944, 25–28; "On The Trial of an Epidemic: Health Squad Traces Ohio Polio," *Life* magazine, October 23, 1950, 40–41; Roger A. MacDonald, *A Country Doctor's Chronicle: Further Tales from the North Woods* (St. Paul: Minnesota Historical Society Press, 2004), 28, 29, 174.
13. Roger A. Rosenblatt and L. Gary Hart, "Physicians and Rural America," *Western Journal of Medicine* 173, 5 (November 2000): 348–351.
14. "Proceedings of the Committee," May 17, 1951, Box 244, Folder 4, Dr. Jonas E. Salk Papers, MSCL, 33.
15. Ibid., 29–30.
16. "Meeting on Gamma Globulin, Waldorf Astoria, New York City," February 3, 1950, Henry Weiner, Accurate Reporting, Gamma Globulin (Round Table Conference), S 3: Conferences, Box 5, Conferences and Meetings Records, MDA, 162.
17. Rosemary Gibson and Janardan Prasad Singh, *Wall of Silence: The Untold Story of the Medical Mistakes That Kill and Injure Millions of Americans* (Washington, D.C.: LifeLine Press, 2003); Allen M. Hornblum, *Sentenced to Science: One Black Man's Story of Imprisonment in America* (College Park: Pennsylvania State University Press, 2007).
18. James H. Jones, *Bad Blood: The Tuskegee Syphilis Experiment* (New York: The Free Press, 1981); Susan Reverby, *Examining Tuskegee* (Chapel Hill: University of North Carolina Press, 2009); Henry Samuel, "French Bread Spiked with LSD

in CIA Experiment," *The Telegraph,* March 11, 2010, http://www.telegraph.co.uk/news/worldnews/europe/france/7415082/French-bread-spiked-with-LSD-in-CIA-experiment.html, last viewed December 10, 2010; H. P. Albarelli, *A Terrible Mistake: The Murder of Frank Olson and the CIA's Secret Cold War Experiments* (Walterville, Ore.: Trine Day, 2008).

19. Hammon et al., "Evaluation of Red Cross Gamma Globulin," Part 1, 747.
20. Marc Schell, *Polio and Its Aftermath* (Cambridge, Mass.: Harvard University Press, 2005), 202; *Miracle at Hickory* (New York: NFIP, 1944), MDA.
21. "Background Story," 1951, S 3: GG FT, Box 3, SSR, MDA, 4.
22. Hammon et al., "Evaluation of Red Cross Gamma Globulin," Part 1, 744–745.
23. "Proceedings of the Committee," July 6, 1951, Box 244, Folder 4, Dr. Jonas E. Salk Papers, MSCL, 23–24.
24. Ibid., 29.
25. "Proceedings of the Committee," May 17, 1951, Box 244, Folder 4, Dr. Jonas E. Salk Papers, MSCL, 20.
26. Ibid., 18.
27. David L. Sills, *The Volunteers: Means and Ends in a National Organization* (Glencoe, Ill.: Free Press, 1957), 28, 40, 165–167.
28. "Polio Precautions," *The Corsicana* (Texas, Semi-Weekly Light), April 18, 1950; "Polio Pointers," *Chillicothe Constitution-Tribune* (Chillicothe, Missouri), September 6, 1951, 3.
29. "Frame Rules to Check Infantile Paralysis," *New York Times,* July 29, 1944, 10; "Rules to Prevent Paralysis Given," *New York Times,* June 1, 1949, 41; "Five Polio Precautions Are Listed for Parents," *Emmetsburg (Iowa) Reporter,* April 12, 1949, 2.
30. Alton L. Blakeslee, "Doctors Advise: Postpone Vaccinations for Children during Warm Polio Months," *Dixon Evening Telegraph,* May 29, 1951, 9.
31. "Payson Youth Gets Final Serum Shot," *Deseret News,* September 8, 1951, 5; Charles R. Rinaldo, "Passive Immunization against Poliomyelitis: The Hammon Gamma Globulin Field Trials, 1951–1953," *AJPH* 95, 5 (May 2005): 793.
32. Harry Marks, *The Progress of Experiment: Science and Therapeutic Reform in the United States, 1900–1990* (Cambridge: Cambridge University Press, 1997, 2000), 136–137, 139.
33. "Medicine: Closing In on Polio," *Time,* March 29, 1954; Ducas to O'Connor, July 3, 1951, S 14: Polio, Box 12, MPR, MDA.
34. Allen M. Hornblum, *Acres of Skin: Human Experiments at Holmesburg Prison* (New York: Routledge, 1998); Jordan Goodman et al., eds., *Useful Bodies: Humans in the Service of Medical Science in the Twentieth Century* (Baltimore: Johns Hopkins University Press, 2003).
35. "Proceedings of the Committee," May 17, 1951, Box 244, Folder 4, Dr. Jonas E. Salk Papers, MSCL, 18.
36. "Ducas, Dorothy," uncorrected version, March 26, 1984, Box 1, Oral History Records (henceforth OHR), MDA, 2.
37. Dorothy Ducas, "In Miniature: Mrs. Charles H. Sabin," *McCall's,* September 1930, 4.
38. "Ducas, Dorothy," March 26, 1984, Box 1, OHR, MDA, 2.
39. Scott M. Cutlip, *Fund Raising in the United States: Its Role in America's Philanthropy* (Piscataway, N.J.: Transaction Publishers, 1990), 384.
40. "Ducas, Dorothy," March 26, 1984, Box 1, OHR, MDA, 2.
41. "Women Open 'Week' to Fight Paralysis," *New York Times,* January 20, 1942, 16.

42. Allan M. Winkler, *The Politics of Propaganda: The Office of War Information, 1942–1945* (New Haven: Yale University Press, 1978).
43. Kenneth Paul O'Brien and Lynn H. Parsons, *The Home-Front War: World War II and American Society* (Westport, Conn.: Greenwood Publishing Group, 1995), 89–91; "Ducas, Dorothy," March 26, 1984, Box 1, OHR, MDA, 12; Maureen Honey, *Creating Rosie the Riveter: Class, Gender, and Propaganda during World War II* (Amherst: University of Massachusetts Press, 1984), 37–39.
44. "Polio Foundation Names Chief of Public Relations," *New York Times*, November 28, 1949, 14.
45. Weaver to White, August 6, 1951, S 3: GG FT, Box 3, SSR, MDA.
46. Samuel H. Williamson, "Seven Ways to Compute the Relative Value of a U.S. Dollar Amount, 1790 to Present," Measuring Worth, 2010, http://www.measuringworth.com/uscompare/, last viewed September 3, 2010. CPI in 2010 was $826,000.
47. L. Pearce Williams, "Chapter VIII: Passive Immunity to Poliomyelitis: The Field Testing of Gamma Globulin as a Prophylactic Agent," Preliminary Draft, S 1: History, Box 1, History of the NFIP Records (henceforth HNR), MDA, 45.
48. Hammon to Weaver, July 31, 1951, S 3: GG FT, Box 3, SSR, MDA.
49. Jane S. Smith, *Patenting the Sun: Polio and the Salk Vaccine* (New York: William Morrow, 1990), 213–214; Jim Hartz and Rick Chappell, *Worlds Apart: How the Distance between Science and Journalism Threatens America's Future* (Nashville: First Amendment Center, 1997), http://www.freedomforum.org/publications/first/worldsapart/worldsapart.pdf, last viewed October 12, 2010.
50. Williams, "Chapter VIII: Passive Immunity," Preliminary Draft, S 1: History, Box 1, HNR, MDA, 46.
51. William McD. Hammon, "Suggested Plans for a Field Trial to Determine the Effectiveness of Gamma Globulin in the Prophylaxis of Poliomyelitis," April 1951, Folder 4, Box 244, Dr. Jonas E. Salk Papers (1926–1991), Mandeville Special Collections Library, La Jolla, California, 10. See also "Proceedings of the Committee," July 6, 1951, 30.
52. Williams, "Chapter VIII: Passive Immunity," Preliminary Draft, S 1: History, Box 1, HNR, MDA, 10; "Epidemic Hits Houston," *San Antonio Light*, June 18, 1948, 2-A.
53. Champlin to Kumm, August 28, 1951, S 14: Polio, Box 12, MPR, MDA.
54. "Henry W. Kumm, 89, Foundation Official," *New York Times*, January 15, 1991, D 19; Kumm, Henry W., Biographical Data, Series: Human Resources, Box 9, MPR, MDA.
55. Weaver to Sullivan, August 15, 1951, S 3: GG FT, Box 3, SSR, MDA.
56. Weaver to Hammon, August 21, 1951, S 3: ibid.; Ducas to Weaver, July 23, 1951, "Background Story," August 1951, S 14: Polio, Box 12, MPR, MDA.
57. Weaver to Hammon, August 21, 1951, S 3: GG FT, Box 3, SSR, MDA, 2.
58. Van Riper to O'Connor, June 21, 1951, S 14: Polio, Box 12, MPR, MDA.
59. Ruth Horowitz, *In the Public Interest: Medical Licensing and the Disciplinary Process* (New Brunswick, N.J.: Rutgers University Press, 2012), chapter 2; John Duffy, *From Humors to Medical Science: A History of American Medicine* (Urbana: University of Illinois Press, 1993), chapters 9 and 20.
60. Ryan to White, July 25, 1951, S 14: Polio, Box 12, MPR, MDA; Ryan to Martin, July 25, 1951, S 14: ibid.
61. "Martin Clearwater and Bell LLP, Firm History," http://www.mcblaw.com/index.php?option=content&task=view&id=21&Itemid=34, last viewed January 30, 2009.

62. Weaver to Van Riper, July 31, 1951, S 3: GG FT, Box 3, SSR, MDA; Martin to Ryan, 7 August 1951, S 3: ibid.
63. Martin to Ryan, August 7, 1951, S 3: ibid.
64. Walter Lippman, *Public Opinion* (New York: Macmillan, 1922), chapter 15, "Leaders and the Rank and File," section 4.
65. NFIP, "Proposal for a Field Study of the Value of Gamma Globulin as a Prophylactic Agent in Poliomyelitis," July 25, 1944, S 14: Polio, Box 12, MPR, MDA, 4.
66. George S. Day, "Strategic Market Analysis: Top-Down and Bottom-Up Approaches," *Marketing Science Institute* (1980): 80–105, http://www.msi.org/publications/publication.cfm?pub=110, last viewed August 14, 2011.
67. Weaver to Hammon, July 25, 1951, S 3: GG FT, Box 3, SSR, MDA.
68. Martin to Ryan, August 7, 1951, S 3: ibid.
69. Paul V. Dutton, *Differential Diagnosis: A Comparative History of Health Care Problems and Solutions in the United States and France* (Ithaca, N.Y.: Cornell University Press, 2007); Paul Starr, *The Social Transformation of American Medicine* (New York: Basic Books, 1982), book 2, chapter 3.
70. Rima M. Apple, *Perfect Motherhood: Science and Childrearing in America* (New Brunswick, N.J.: Rutgers University Press, 2006), chapters 3 and 4.
71. Weaver to Hammon, July 25, 1951, S 3: GG FT, Box 3, SSR, MDA.
72. Susan E. Lederer, *Subjected to Science: Human Experimentation in America before the Second World War* (Baltimore: Johns Hopkins University Press, 1997).
73. Vivien Spitz, *Doctors from Hell: The Horrific Account of Nazi Experiments on Humans* (Boulder, Colo.: Sentient Publications, 2005), 19–25, 253–255.
74. James Jones, *Bad Blood*; Keith Wailoo, *Dying in the City of the Blues: Sickle Cell Anemia and the Politics of Race and Health* (Chapel Hill: University of North Carolina Press, 2001); Andrew Scull, *Madhouse: A Tragic Tale of Megalomania and Modern Medicine* (New Haven: Yale University Press, 2005); Sydney Ann Halpern, *Lesser Harms: The Morality of Risk in Medical Research* (Chicago: University of Chicago Press, 2006); Jordan Goodman et al., eds., *Useful Bodies*; Hornblum, *Acres of Skin*; Jonathan D. Moreno, *Undue Risk: Secret State Experiments on Humans* (New York: Routledge, 2001); Hans Jonas, "Philosophical Reflections on Experimenting with Human Subjects," in *Experimentation with Human Subjects*, ed. P. A. Freund (New York: George Braziller, 1970), 1–31.
75. Hammon et al., "Evaluation of Red Cross Gamma Globulin," Part 1, 744.
76. Lederer, *Subjected to Science*, 4; W. K. Jaques, "Experiments and Observations in Scarlet Fever," *JAMA* 34 (1900): 1302.
77. Hammon et al., "Evaluation of Red Cross Gamma Globulin," Part 1, 744.
78. Ducas to O'Connor, July 3, 1951, S 14: Polio, Box 12, MPR, MDA.
79. Weaver to Hammon, August 21, 1951, S 3: GG FT, Box 3, SSR, MDA, 2.
80. "Background Story," August 1951, S 3: ibid., 5.
81. "News Story," August 1951, S 14: GG FT, Box 12, MPR, MDA, 1.
82. "Radio," August 1951, S 14: ibid.
83. Van Riper to O'Connor, June 21, 1951, S 14: Polio, Box 12, MPR, MDA.
84. "Proceedings of the Committee on Immunization," May 17, 1951, 14.
85. Ducas to O'Connor, July 3, 1951, S 14: Polio, Box 12, MPR, MDA.
86. Martha Stephens, *The Treatment: The Story of Those Who Died in the Cincinnati Radiation Tests* (Durham, N.C.: Duke University Press, 2002), 103.
87. Weaver to Van Riper, July 18, 1951, S 14: Polio, Box 12, MPR, MDA.

88. Ducas to O'Connor, July 3, 1951, S 14: ibid., 2; Weaver to Van Riper, July 18, 1951, S 14: ibid.
89. Ducas to O'Connor, July 3, 1951, S 14: ibid.
90. Weaver to Van Riper, July 18, 1951, S 14: ibid.
91. St. John to All Writers, August 29, 1951, S 14: ibid.
92. Lewis G. Decker, *Images of America: Johnstown* (Charleston, S.C.: Arcadia Publishing, 1999), 37–50. See also Hammon et al., "Evaluation of Red Cross Gamma Globulin," Part 1, 746.
93. Hammon to Ashworth, August 10, 1951, S 3: GG FT, Box 3, SSR, MDA; E. Pickering, "American Chemical Industries—E. R. Squibb & Sons," *Industrial & Engineering Chemistry* 22, 6 (June 1930): 682–684.
94. Douglas Starr, *Blood: An Epic History of Medicine and Commerce* (New York: Alfred Knopf, 1998), 113; Ashworth to Hammon, November 5, 1951, S 3: GG FT, Box 3, SSR, MDA.
95. Allan M. Brandt, "Polio, Politics, Publicity, and Duplicity: Ethical Aspects in the Development of the Salk Vaccine," *International Journal of Health Services* 8, 2 (1978): 257–270; Anita Guerrini, *Experimenting with Humans and Animals: From Galen to Animal Rights* (Baltimore: Johns Hopkins University Press, 2003), chapter 6.
96. Victoria A. Harden, "A Short History of the National Institutes of Health," Office of NIH History, http://history.nih.gov/exhibits/history/, last viewed August 5, 2010; "World War II and the Postwar Period: Standards for Essential Blood Products," *CBER Vision* (Center for Biologics Evaluation and Research, July 2002), 7–8.
97. Kumm to Hammon, August 10, 1951, S 3: GG FT, Box 3, SSR, MDA.
98. T. Nakayama and C. Aizawa, "Change in Gelatin Content of Vaccines Associated with Reduction in Reports of Allergic Reactions," *Journal of Allergy and Clinical Immunology* 106 (2000): 591–592; M. Sakaguchi, T. Nakayama, and S. Inouye, "Food Allergy to Gelatin in Children with Systemic Immediate-Type Reactions, Including Anaphylaxis, to Vaccines," *Journal of Allergy and Clinical Immunology* 98 (1996): 1058–1061.
99. D. D. Kozoll, H. Popper, F. Steigmann, and B. W. Volk, "The Use of Gelatin Solutions in the Treatment of Human Shock," *American Journal of the Medical Sciences* 208 (1944): 141.
100. Kumm to Hammon, August 15, 1951, S 3: GG FT, Box 3, SSR, MDA.
101. Thomas P. Hughes, *American Genesis: A Century of Invention and Technological Enthusiasm, 1870–1970* (Chicago: University of Chicago Press, 2004), 2–8, 184–185.
102. Kumm to Hammon, August 14, 1951, S 3: GG FT, Box 3, SSR, MDA.
103. Kumm to Hammon, August 15, 1951, S 3: ibid.
104. "Sam T. Gibson, 83, Dies; Was Director of National Blood Bank for Red Cross," *Washington Post*, September 25, 1999; Sam T. Gibson, "Gamma Globulin," *American Journal of Nursing* 53, 6 (June 1953): 700–703; Gibson to Hammon, December 27, 1951, S 3: GG FT, Box 3, SSR, MDA.
105. Kumm to Hammon, August 10, 1951, S 3: GG FT, Box 3, SSR, MDA.
106. Kumm to Hammon, August 15, 1951, S 3: ibid.
107. Hammon to Ashworth, August 10, 1951, S 3: ibid.
108. Hammon to Ashworth, August 10, 1951, S 3: ibid; Kumm to Hammon, August 24, 1951, S 3: ibid.
109. Kumm to Hammon, August 10, 1951, S 3: ibid.
110. Hammon to Ashworth, August 10, 1951, S 3: ibid., 1.

111. Kumm to Hammon, August 14, 1951, S 3: ibid., 1–2.
112. Kumm to Hammon, August 24, 1951, S 3: GG FT, Box 3, SSR, MDA.
113. Hammon to Ashworth, August 10, 1951, S 3: ibid., 1–2.
114. Kumm to Hammon, August 24, 1951, S 3: ibid.
115. Hammon et al., "Evaluation of Red Cross Gamma Globulin," Part 1, 742; Marks, *Progress of Experiment*, 141–148.
116. W. A. Silverman, "Personal Reflections on Lessons Learned from Randomized Trials Involving Newborn Infants, 1951 to 1967," interview by Iain Chalmers in October 2003, James Lind Library, http://www.jameslindlibrary.org/essays/cautionary/silverman.html, last viewed December 10, 2010.
117. Susan L. Smith and Stephen Mawdsley, "Alberta Advantage: A Canadian Proving Ground for American Medical Research on Mustard Gas and Polio in the 1940s and 50s," in *Locating Health*, ed. Erika Dyck and Christopher Fletcher (London: Pickering and Chatto, 2011), 102.
118. Hammon et al., "Evaluation of Red Cross Gamma Globulin," Part 1, 744.
119. Champlin to Hammon, August 24, 1951, S 3: GG FT, Box 3, SSR, MDA; Hammon to Kumm, November 30, 1951, S 3: ibid., 2.
120. R. Duncan Luce and Howard Raiffa, *Games and Decisions: Introduction and Critical Survey* (New York: Wiley, 1957).
121. Weaver to Hemphill, July 17, 1951, S 3: GG FT, Box 3, SSR, MDA.
122. Alexander G. Gilliam, Fay M. Hemphill, and Jean H. Gerende, "Poliomyelitis Epidemic Recurrence in the Counties of the United States, 1932–1946," *Public Health Reports* 64, 49 (December 9, 1949): 1595; F. M. Hemphill, "Methods of Predicting Total Cases of Poliomyelitis during Epidemic Periods," *AJPH* 42 (August 1952): 947–955, 947.
123. Hemphill, "Methods of Predicting Total Cases of Poliomyelitis," 948.
124. Hammon to Kumm, November 30, 1951, S 3: GG FT, Box 3, SSR, MDA, 2.
125. Anahad O'Connor, "Lewis L. Coriell, 90, Virologist Who Set Stage for Polio Vaccine," *New York Times*, July 2, 2001; "Biographical Sketches of Dr. William McD. Hammon and Assistants," 1951, S 3: GG FT, Box 3, SSR, MDA, 7.
126. "Report of Gamma Globulin Passive Immunization Field Project, Utah County, Utah, 1951," November 30, 1951, S 14: Polio, Box 12, MPR, MDA.
127. Lyons to Officials, "Research Project with Gamma Globulin," August 27, 1951, S 14: ibid., 1.
128. "Polio Planners Favor Starting School as Usual," *Greeley Daily Tribune* (Colorado), August 22, 1951, 13.
129. "Report of Gamma Globulin Passive Immunization Field Project, Utah County, Utah, 1951," November 30, 1951, S 14: Polio, Box 12, MPR, MDA, 1–2.
130. William C. Patrick, "Provo Site for Study of Polio," *Salt Lake Tribune*, September 1, 1951, 17.
131. Census of Population—1950—Volume II, Part 44, Characteristics of the Population: Utah, http://www2.census.gov/prod2/decennial/documents/37784400v2p44ch3.pdf, last viewed December 10, 2010.
132. Pamela S. Perlich, *Utah Minorities: The Story Told by 150 Years of Census Data* (Bureau of Economic and Business Research, David S. Eccles School of Business, University of Utah, October 2002); "Table 41.—General Characteristics of the Population, For Counties: 1950," Utah County, http://www2.census.gov/prod2/decennial/documents/37784400v2p44ch3.pdf, last viewed January 24, 2016.
133. Becker to Hanks, August 30, 1951, S 14: Polio, Box 12, MPR, MDA.

134. Kumm to Hammon, August 24, 1951, S 3: GG FT, Box 3, SSR, MDA, 1–2. For information about weight, see "Diary of Dr. Henry W. Kumm," S 3: ibid., 1.
135. "Diary of Dr. Henry W. Kumm," S 3: ibid.
136. Lyons to Officials, "Research Project," August 27, 1951, S 14: Polio, Box 12, MPR, MDA, 1.
137. Rinaldo, "Passive Immunization," 794; Hammon et al., "Evaluation of Red Cross Gamma Globulin," Part 1, 747.
138. Lyons to Officials, "Research Project," August 27, 1951, GG Program, S 14: Polio, Box 12, MPR, MDA.
139. Becker to Hanks, August 30, 1951, S 14: ibid.
140. Lyons to Officials, August 27, 1951, S 14: ibid.
141. Suchomel to Weaver, August 8, 1951, S 3: GG FT, Box 3, SSR, MDA.

Chapter 4 – The Pilot Study

1. "Drama of Human Welfare: Vital Polio Experiments Begin Today in Provo," *Salt Lake Tribune*, September 4, 1951, 17.
2. "First Child," *Provo Daily Herald*, September 4, 1951, 3.
3. "County Medics Plan to Install President," *Deseret News*, December 5, 1950, 8A.
4. "Diary of Dr. Henry W. Kumm," August 1951, S 3: GG FT, Box 3, Surveys and Studies Records (henceforth SSR), March of Dimes Archives (henceforth MDA).
5. Weaver to Hammon, July 25, 1951, GG FT, Box 3, SSR, MDA, 1; "Diary of Dr. Henry W. Kumm," August-September 1951, ibid.
6. William McD. Hammon, Lewis L. Coriell, and Joseph Stokes, "A Series of Controlled Field Tests to Evaluate Red Cross Gamma Globulin as a Prophylactic for Poliomyelitis: 1. Plan of Tests, Conduct, and Results of 1951 Pilot Study in Utah," 1951, S 14: Polio, Box 12, Medical Program Records (henceforth MPR), MDA, 20.
7. "Diary of Dr. Henry W. Kumm," August-September 1951, GG FT, Box 3, SSR, MDA, 1.
8. Ibid.
9. "Proceedings of the Committee on Immunization of the National Foundation for Infantile Paralysis," December 4, 1951, Folder 6, Box 244, Dr. Jonas E. Salk Papers (1926–1991), Mandeville Special Collections Library (henceforth MSCL), La Jolla, California, 5–6.
10. William McD. Hammon, "Address to the Medical Society," August 1951, S 3: GG FT, Box 3, SSR, MDA, 4.
11. Hammon, "Address to the Medical Society," ibid., 9.
12. R. L. Kane, "Medical Imperialism," *Journal of Community Health* 5, 2 (December 1979): 81; Ann Murcott, ed., *Sociology and Medicine* (Burlington, Vt.: Ashgate Publishing, 2006), chapter 5.
13. Hammon, "Address to the Medical Society," August 1951, S 3: GG FT, Box 3, SSR, MDA, 1–9.
14. Ibid., 9.
15. Ibid., 7–8.
16. William C. Patrick, "Provo Site for Study of Polio," *Salt Lake Tribune*, September 1, 1951, 17.
17. Jonathan D. Moreno, ed., *In the Wake of Terror: Medicine and Morality in a Time of Crisis* (Cambridge, Mass.: MIT Press, 2003), chapter 1.
18. K. A. Cuordileone, *Manhood and American Political Culture in the Cold War: Masculinity, the Vital Center, and American Political Culture in the Cold War, 1949–1963* (New York: Routledge, 2005).

19. Kristine Hammond, interview by Stephen E. Mawdsley, October 16, 2009, corrected version, 4–5.
20. Daniel A. Menchik and David O. Meltzer, "The Cultivation of Esteem and Retrieval of Scientific Knowledge in Physician Networks," *Journal of Health and Social Behavior* 51, 2 (June 2010): 137–152.
21. "Obituary: Dr. Roy B. Hammond," *Deseret News*, March 25, 2004, http://www.deseretnews.com/article/1247488/Obituary-Dr-Roy-B-Hammond.html, last viewed May 12, 2011.
22. Reed Smoot served as a United States senator from 1903 to 1932 and a member of the governing Council of Twelve Apostles of the Church of Jesus Christ of Latter Day Saints from 1898 to 1941. His father, Abraham Owen Smoot, was a LDS missionary, pioneer, and the second mayor of Salt Lake City. See "Abraham Owen Smoot," http://www.media.utah.edu/UHE/s/SMOOT,ABRAHAM.html, last viewed August 17, 2009; "The Congress: Spending Spree," *Time*, April 8, 1940; See also "Reed Smoot," Brigham Young University High School, http://www.byhigh.org/History/Smoot/Reed.html, last viewed August 17, 2009.
23. "Diary of Dr. Henry W. Kumm," September 1951, S 3: GG FT, Box 3, SSR, MDA, 1.
24. Lee B. Kennett, *G.I.: The American Soldier in World War II* (New York: Scribner, 1987), 139, chap. 12; G. H. Elder Jr. and E. C. Clipp, "Wartime Losses and Social Bonding: Influences across 40 Years in Men's Lives," *Psychiatry* 51, 2 (May 1988): 177–198.
25. Mary Ann Clawson, *Constructing Brotherhood: Class, Gender, and Fraternalism* (Princeton: Princeton University Press, 1989).
26. Hammond, interview, 6–8.
27. "Diary of Dr. Henry W. Kumm," September 1951, S 3: GG FT, Box 3, SSR, MDA, 1; W. M. Hammon, L. L. Coriell, and J. Stokes Jr., "Evaluation of Red Cross Gamma Globulin as a Prophylactic Agent for Poliomyelitis: Part 1. Plan of Controlled Field Tests and Results of the 1951 Pilot Study in Utah," *JAMA* 150 (1952): 747.
28. "Diary of Dr. Henry W. Kumm," August/September 1951, S 3: GG FT, Box 3, SSR, MDA.
29. Hammon et al., "Evaluation of Red Cross Gamma Globulin," Part 1, 745.
30. "Diary of Dr. Henry W. Kumm," August-September 1951, S 3: GG FT, Box 3, SSR, MDA.
31. Leon N. Perry, "Utah Waits Thirty Polio Researchers," *Deseret News*, September 4, 1951, 4B.
32. Susan J. Douglas, *Listening In: Radio and the American Imagination* (New York: Times Books, 1999).
33. Photograph #51–984, August/September 1951, S 2: GG FT, Box 3, Photographic Records, MDA.
34. Photograph #51–920, August/September 1951, S 2: ibid.
35. "Diary of Dr. Henry W. Kumm," August-September 1951, S 3: GG FT, Box 3, SSR, MDA, 2.
36. St. John to Weaver, April 21, 1952, S 14: Polio, Box 12, MPR, MDA, 2.
37. "Diary of Dr. Henry W. Kumm," August-September 1951, S 3: GG FT, Box 3, SSR, MDA, 2.
38. "Utah County Chosen for First Mass Tests against Polio," *Salt Lake Telegram*, August 31, 1951, 1.
39. Leon N. Perry, "Utah Waits Thirty Polio Researchers," *Deseret News*, September 4, 1951, 1B.

40. Sydney A. Halpern, *Lesser Harms: The Morality of Risk in Medical Research* (Chicago: University of Chicago Press, 2006), 128.
41. Ibid.
42. "Diary of Dr. Henry W. Kumm," August-September 1951, S 3: GG FT, Box 3, SSR, MDA, 3.
43. "Utah County Opens Polio Clinics to Inoculate 5000 Children," *Salt Lake Telegram*, September 4, 1951, 1.
44. Andrew Scull, *Madhouse: A Tragic Tale of Megalomania and Modern Medicine* (New Haven: Yale University Press, 2005), 58.
45. Richard N. Ostling and Joan K. Ostling, *Mormon America: The Power and the Promise* (New York: HarperOne, 2007), 171–172.
46. Hammon et al., "Evaluation of Red Cross Gamma Globulin," Part 1, 744.
47. William McD. Hammon, "Report of Gamma Globulin Passive Immunization Field Project, Utah County, Utah, 1951," November 30, 1951, S 14: Polio, Box 12, MPR, MDA, 4.
48. Theron H. Luke, "Eyes of U.S. on Utah County Anti-Polio Test," *Provo Daily Herald*, September 2, 1951, 2.
49. "Diary of Dr. Henry W. Kumm," September 1, 1951, S 3: GG FT, Box 3, SSR, MDA.
50. Ostling and Ostling, *Mormon America*, 159.
51. Hammond, interview, 12.
52. "Diary of Dr. Henry W. Kumm," August-September 1951, S 3: GG FT, Box 3, SSR, MDA, 2.
53. Perry, "Utah Waits Thirty Polio Researchers," 4B.
54. "Stake Heads Ask Polio Test Backing," *Provo Daily Herald*, September 4, 1951.
55. Jonathan B. Imber, *Trusting Doctors: The Decline of Moral Authority in American Medicine* (Princeton: Princeton University Press, 2008), chapter 4.
56. "Stake Heads Ask Polio Test Backing," *Provo Daily Herald*, September 4, 1951.
57. "Poliomyelitis Serum Clinics, [Read in Utah County Churches, 2 September 1951]," September 1951, S 14: Polio, Box 12, MPR, MDA.
58. Ostling and Ostling, *Mormon America*, 216–221.
59. Hammon, "Report of Gamma Globulin," November 30, 1951, S 14: Polio, Box 12, MPR, MDA, 3; Perry, "Utah Waits Thirty Polio Researchers," 1B.
60. "Polio Claims Another in Utah County," *Salt Lake Tribune*, September 7, 1951, 13.
61. "News Story," September 7, 1951, S 14: Polio, Box 12, MPR, MDA.
62. "Provo Mother of Polo-Stricken Children Asks Support of Test," *Provo Daily Herald*, September 4, 1951, 3.
63. Ibid.
64. James Jex, interview by Stephen E. Mawdsley, November 3, 2009, corrected version, 2.
65. "Proceedings of the Committee," December 4, 1951, Folder 6, Box 244, Dr. Jonas E. Salk Papers (1926–1991), MSCL, 8.
66. Perry, "Utah Waits Thirty Polio Researchers," 4B.
67. Halpern, *Lesser Harms*, 4.
68. Ibid., 5.
69. Perry, "Utah Waits Thirty Polio Researchers," 4B.
70. "News Story," September 5, 1951, S 14: Polio, Box 12, MPR, MDA.
71. Rima M. Apple, *Perfect Motherhood: Science and Childrearing in America* (New Brunswick, N.J.: Rutgers University Press, 2006), 83–90.
72. "Test Immunizing to Continue at Quincy School," *Ogden Examiner*, February 20, 1951, 3.

73. William C. Patrick, "State Defense Aid Urges Medical Preparedness," *Salt Lake Tribune*, October 28, 1950, 17.
74. "City-Wide Health Program of Immunization Billed," *Salt Lake Tribune*, November 26, 1950, 16A.
75. "Good Sense Sends Children to Immunization Clinics," *Salt Lake Tribune*, December 2, 1950, 15.
76. "Tooele School Pupils Get Special Shots," *Salt Lake Tribune*, October 24, 1950, 5; "P-TA Urges Immunization of Children," *Salt Lake Tribune*, December 6, 1950, 18.
77. William C. Patrick, "1459 Utah Children Inoculated in Provo Polio Experiment," *Salt Lake Tribune*, September 5, 1951, 15.
78. Hammon, "Report of Gamma Globulin," November 30, 1951, S 14: Polio, Box 12, MPR, MDA, 3.
79. Becker to Weaver, October 11, 1951, S 14: Polio, Box 12, MPR, MDA.
80. Ibid.
81. James Colgrove, *State of Immunity: The Politics of Vaccination in Twentieth-Century America* (Berkeley: University of California Press, 2006), chapter 4.
82. Patrick, "1459 Utah Children Inoculated in Provo Polio Experiment," 15.
83. Hammon et al., "Evaluation of Red Cross Gamma Globulin," Part 1, 748.
84. Photograph #51–1001, S 2: NFIP, GG Field Trial, Provo, Utah, Box 3, Photographic Records, MDA.
85. "News Story," September 5, 1951, S 14: Polio, Box 12, MPR, MDA.
86. "Polio Serum Clinics Cheer Support in Utah County," *Salt Lake Tribune*, September 7, 1951, 13.
87. Peter Morrall and Mike Hazelton, "Architecture Signifying Social Control: The Restoration of Asylumdom in Mental Health Care?," *Australian and New Zealand Journal of Mental Health Nursing* 9, 2 (2000): 89–96; Michel Foucault, *Discipline and Punish* (New York: Vintage Books, 1995), part 3, chapter 3.
88. Allen Kent Powell, ed., "Spanish Fork," in *Utah History Encyclopedia* (Salt Lake City: University of Utah Press, 1994), http://www.media.utah.edu/UHE/UHEindex.html, last viewed March 5, 2010.
89. Pierce and Hammon, "Digest of Duties and Functions," June 6, 1952, S 3: GGFT, Box 3, SSR, MDA, 21–26. For evidence of uniformed policemen, see *Operation Marbles and Lollipops*, NFIP (DVD, 1952; New York: MDA, 2009).
90. Pierce and Hammon, "Digest of Duties and Functions," June 6, 1952, S 3: GGFT, Box 3, SSR, MDA, 21.
91. David M. Oshinsky, *A Conspiracy So Immense: The World of Joe McCarthy* (New York: Oxford University Press, 2005); Elaine Tyler May, *Homeward Bound: American Families in the Cold War* (New York: Basic Books, 1990); Foucault, *Discipline and Punish*, 202.
92. Hammon et al., "A Series of Controlled Field Tests," S 14: Polio, Box 12, MPR, MDA, 13.
93. Hammon, "Report of Gamma Globulin," November 30, 1951, S 14: ibid., 4.
94. Pierce and Hammon, "Digest of Duties and Functions," June 6, 1952, S 3: GGFT, Box 3, SSR, MDA, 22.
95. William Hammon and Robert Pierce, "Diagram: Station #5, Injection Room," June 6, 1952, S 3: ibid., 25.
96. T. Kushner, "Doctor-Patient Relationships in General Practice—a Different Model," *Journal of Medical Ethics* 7 (1981): 128–131.
97. "News Story," September 6, 1951, GG Program, S 14: Polio, Box 12, MPR, MDA.

98. "News Story," September 5, 1951, GG Program, S 14: ibid.
99. Perry, "Utah Waits Thirty Polio Researchers," 4B.
100. Hammon, "Report of Gamma Globulin," November 30, 1951, S 14: Polio, Box 12, MPR, MDA, 3.
101. Hammon to Kumm, October 13, 1951, S 3: GG FT, Box 3, SSR, MDA, 2.
102. Ibid.
103. Jex, interview, 3.
104. K. V. Iserson, "The Origins of the Gauge System for Medical Equipment," *Journal of Emergency Medicine* 5 (1987): 45–48.
105. F. Ralph Berberich and Zachary Landman, "Reducing Immunization Discomfort in 4- to 6-Year-Old Children: A Randomized Clinical Trial," *Pediatrics* 124, 2 (2009): 203–209.
106. Perry, "Utah Waits Thirty Polio Researchers," 4B.
107. Jex, interview, 4.
108. Perry, "Utah Waits Thirty Polio Researchers," 4B.
109. "Project Needle-Lollipop," *Life* magazine, September 24, 1951, 71.
110. Photograph #51–928, Inoculations, GG, S 2: NFIP, Box 3, Photographic Records, MDA.
111. "News Story," September 5, 1951, S 14: Polio, Box 12, MPR, MDA.
112. Patrick, "1459 Utah Children Inoculated in Provo Polio Experiment," 15.
113. Perry, "Utah Waits Thirty Polio Researchers," 4B.
114. Photograph, #51–972A, Provo, Utah, GG Field Test, S 2, Box 3, Photographic Records, MDA.
115. Chappell to Ducas and Stegen, July 3, 1952, S 14: Polio, Box 12, MPR, MDA.
116. Hammon to Kumm, September 13, 1951, S 3: GG FT, Box 3, SSR, MDA.
117. Paul Thomas Young, *Motivation of Behavior: The Fundamental Determinants of Human and Animal Activity* (New York: J. Wiley & Sons., 1936), 33–34.
118. Hammon, "Report of Gamma Globulin," November 30, 1951, S 14: Polio, Box 12, MPR, MDA, 4.
119. "Medics Close Polio Test, Wait Results," *Salt Lake Tribune,* September 8, 1951, 14.
120. "News Story," September 7, 1951, S 14: Polio, Box 12, MPR, MDA; "Polio Testing Enters County," *Deseret News*, September 6, 1951, 1A.
121. Hammon, "Report of Gamma Globulin," November 30, 1951, S 14: Polio, Box 12, MPR, MDA, 5.
122. Hammon et al., "Evaluation of Red Cross Gamma Globulin," Part 1, 747–748.
123. "Medics Close Polio Test, Wait Results," 14.
124. "Polio Serum Clinics Cheer Support in Utah County," 13; Patrick, "1459 Utah Children Inoculated in Provo Polio Experiment," 15.
125. "Medics Close Polio Test, Wait Results," *Salt Lake Tribune*, 14.
126. William C. Patrick, "How Long Will It Be," *Salt Lake Tribune*, September 9, 1951.
127. Hammon, "Report of Gamma Globulin," November 30, 1951, S 14: Polio, Box 12, MPR, MDA, 7.
128. Ibid., 10.
129. Jessica Wang, *American Science in an Age of Anxiety: Scientists, Anticommunism, and the Cold War* (Chapel Hill: University of North Carolina Press, 1999), 290; Allan A. Needell, *Science, Cold War, and the American State* (New York: Routledge, 2012), 262.
130. Steve Sturdy, "Looking for Trouble: Medical Science and Clinical Practice in the Historiography of Modern Medicine," *Social History of Medicine* (2011) doi:10.1093/shm/hkq106. "Utah County Opens Polio Clinics to Inoculate 5000 Children," 1.

131. James H. Jones, *Bad Blood: The Tuskegee Syphilis Experiment* (New York: Free Press, 1981), 61–77; Susan M. Reverby, ed., *Tuskegee's Truths: Rethinking the Tuskegee Syphilis Study* (Chapel Hill: University of North Carolina Press, 2000); Susan M. Reverby, *Examining Tuskegee* (Chapel Hill: University of North Carolina Press, 2009).
132. Hammon, "Report of Gamma Globulin," November 30, 1951, S 14: Polio, Box 12, MPR, MDA, 6.
133. Hammon to Kumm, September 10, 1951, S 3: GG FT, Box 3, SSR, MDA, 2; Weaver to Barrows, September 24, 1951, S 3: GG FT, Box 3, SSR, MDA.
134. Hammon to Stokes, September 13, 1951, St65p, Hammon, W. McD. #6, Joseph Stokes, Jr., Papers, American Philosophical Society, Philadelphia, Pennsylvania.
135. "Project Needle-Lollipop," *Life* magazine, September 24, 1951, 71–72.
136. Hammon et al., "Evaluation of Red Cross Gamma Globulin," Part 1, 745.
137. Becker to Weaver, October 11, 1951, S 14: Polio, Box 12, MPR, MDA.
138. Hammon, "Report of Gamma Globulin," November 30, 1951, S 14: ibid., 4.
139. Elliott F. Ellis, and Christopher S. Henney, "Adverse Reactions Following Administration of Human Gamma Globulin," *Journal of Allergy* 43, 1 (January 1969): 45–54. See also S. J. Taub, "Adverse Reactions Following Administration of Human Gamma Globulin," *Eye, Ear, Nose, & Throat Monthly* 48, 6 (June 1969): 93–94.
140. Hammon et al., "Evaluation of Red Cross Gamma Globulin," Part 1, 745.
141. Hammon, "Report of Gamma Globulin," November 30, 1951, S 14: Polio, Box 12, MPR, MDA, 6.
142. "Proceedings of the Committee," December 4, 1951, Folder 6, Box 244, Dr. Jonas E. Salk Papers (1926–1991), MSCL, 8.
143. Becker to Weaver, October 11, 1951, S 14: Polio, Box 12, MPR, MDA.
144. W. M. Hammon, L. L. Coriell, P. F. Wehrle, J. Stokes Jr., "Evaluation of Red Cross Gamma Globulin as a Prophylactic Agent for Poliomyelitis: 4. Final Report of Results Based on Clinical Diagnoses," *JAMA* 151 (1953): 1278.
145. Hammon, Draft Reply Card, January 7, 1952, S 3: GG FT, Part I–IX, Box 3, SSR, MDA.
146. Hammon, Coriell, Cheever, Stokes, "Cumulated Incidence of Reported Disease by Week, Month, or Twenty-Week Period after Injection of Gamma Globulin or Gelatin—Final Report," April 2, 1952, S 3: GG FT, Box 3, SSR, MDA.
147. Hammon to Weaver, Kumm, Coriell, Stokes, March 5, 1952, S 3: ibid.
148. Hammon, "Report of Gamma Globulin," S 14: Polio, GG Program, Box 12, MPR, MDA, 9.
149. "Proceedings of the Committee," December 4, 1951, Folder 6, Box 244, Dr. Jonas E. Salk Papers (1926–1991), MSCL, 10–11.
150. Stephen E. Mawdsley, "Balancing Risks: Childhood Inoculations and America's Response to the Provocation of Paralytic Polio," *Social History of Medicine* 26, 4 (November 2013): 759–778.
151. "Proceedings of the Committee," July 6, 1951, Folder 6, Box 244, Dr. Jonas E. Salk Papers (1926–1991), MSCL, 24.
152. Hammon to Kumm, October 13, 1951, S 3: GG FT, Box 3, SSR, MDA.
153. Hammon to Weaver, October 10, 1951, S 3: ibid.
154. "Proceedings of the Committee," December 4, 1951, Folder 6, Box 244, Dr. Jonas E. Salk Papers (1926–1991), MSCL, 13.
155. Birnbaum to Weaver, November 8, 1950, S 3: GG FT, Box 3, SSR, MDA, 1.
156. Hammon to Kumm, October 13, 1951, S 3: GG FT, Box 3, SSR, MDA.

Chapter 5 – Operation Marbles and Lollipops

1. "Polio Testing Nears End; Disease Caused 16th Death," *Sioux City Journal*, July 24, 1952, 8.
2. "Operation Lollipop," *Sioux City Journal*, July 27, 1952, C4.
3. Hammon to Kumm, November 30, 1951, S 3: GG FT, Box 3, Surveys and Studies Records (henceforth SSR), March of Dimes Archives (henceforth MDA).
4. "Proceedings of the Committee on Immunization of the National Foundation for Infantile Paralysis," December 4, 1951, Folder 6, Box 244, Dr. Jonas E. Salk Papers (1926–1991), Mandeville Special Collections Library (henceforth MSCL), La Jolla, California, 8, 12.
5. "Proceedings of the Committee," ibid., 6, 7, 11.
6. Ibid., 11, 12.
7. Ibid., 12, 13.
8. Roland Marchand, *Advertising the American Dream: Making Way for Modernity, 1920–1940* (Berkeley: University of California Press, 1985), 96–100.
9. Allan M. Brandt, *The Cigarette Century* (New York: Basic Books, 2007), 162.
10. Hammond to Hammon, April 4, 1952, St65p, Hammon, W. McD. #16, Joseph Stokes, Jr., Papers, American Philosophical Society, Philadelphia, Pennsylvania (henceforth APS).
11. White to Hammon, March 29, 1952, St65p, Hammon, W. McD. #20, ibid.
12. Utah State Department of Health to Hammon, April 3, 1952, St65p, Hammon, W. McD. #16, ibid.
13. Marchand, *Advertising the American Dream*, 96–100.
14. Hammon to Kumm, September 10, 1951, S 3: GG FT, Box 3, SSR, MDA.
15. D. M. Horstmann, "Poliomyelitis Virus in the Blood of Orally Infected Monkeys and Chimpanzees," *Proceedings of the Society for Experimental Biology and Medicine* 79 (March 1952): 417; Heather A. Carleton, "Putting Together the Pieces of Polio: How Dorothy Horstmann Helped Solve the Puzzle," *Yale Journal of Biology and Medicine* 84, 2 (June 2011): 83–89.
16. D. Bodian, "Pathogenesis of Poliomyelitis in Normal and Passively Immunized Primates after Virus Feeding," *Federation Proceedings* 11 (March 1952): 462; Elizabeth Fee and Manon Parry, "David Bodian," *Proceedings of the American Philosophical Society* 150, 1 (March 2006): 168–172.
17. D. Bodian, "Experimental Studies on Passive Immunizations against Poliomyelitis: The Prophylactic Effect of Human Gamma Globulin on Paralytic Poliomyelitis in Cynomolgus Monkeys after Virus Feeding," *American Journal of Hygiene* 56 (July 1952): 78; W. M. Hammon, L. L. Coriell, and J. Stokes Jr., "Evaluation of Red Cross Gamma Globulin as a Prophylactic Agent for Poliomyelitis: 2. Conduct and Early Followup of 1952 Texas and Iowa-Nebraska Studies," *JAMA* 150 (1952), 751.
18. F. J. Dixon, D. Talmage, and P. H. Maurer, "Half-Life of Homologous Gamma Globulin (Antibody) Molecules in Man, Dog, Rabbit, Guinea Pig, and Mouse," *Federation Proceedings* 11 (March 1952): 466; Hammon et al., "Evaluation of Red Cross Gamma Globulin," Part 2, 750.
19. "Proceedings of the Committee," December 4, 1951, Folder 6, Box 244, Dr. Jonas E. Salk Papers (1926–1991), MSCL, 36.
20. David M. Oshinsky, *Polio: An American Story: The Crusade That Mobilized the Nation against the 20th Century's Most Feared Disease* (Oxford: Oxford University Press, 2005), 157–159.

21. Kumm to Hammon, January 24, 1952, S 3: GG FT, Box 3, SSR, MDA; Robert Pierce and William Hammon, "Digest of Duties and Functions," June 6, 1952, S 3: ibid., 6.
22. Heather Green Wooten, *The Polio Years in Texas: Battling a Terrifying Unknown* (College Station: Texas A&M University Press, 2009), 119–121, 138–142.
23. Charles R. Rinaldo, "Passive Immunization against Poliomyelitis: The Hammon Gamma Globulin Field Trials, 1951–1953," *AJPH* 95, 5 (May 2005): 794; Kumm to Hammon, July 19, 1952, S 3: GG FT, Box 3, SSR, MDA.
24. "Gamma Globulin Passive Immunization Field Project—1952," November 1951, S 3: GG FT, Box 3, SSR, MDA, 1; Losty to Suchomel, March 11, 1952, S 14: Polio, Box 12, Medical Program Records (henceforth MPR), MDA.
25. Hammon et al., "Evaluation of Red Cross Gamma Globulin," Part 2, 752.
26. Chappell to Ducas and Stegen, July 3, 1952, S 14: Polio, Box 12, MPR, MDA.
27. Hammon et al., "Evaluation of Red Cross Gamma Globulin," Part 2, 750.
28. "Suggested Endorsement for Churches and Synagogues," April 1952, S 3: GG FT, Box 3, SSR, MDA; "Field Trial [Radio] Announcements: Group One," April 1952, S 3: ibid.; "Suggested Editorial," April 1952, S 3: ibid.
29. Elmer Bertelsen, "Clinics Are Crowded by Children," *Houston Chronicle*, July 2, 1952, 1.
30. Hammon et al., "Evaluation of Red Cross Gamma Globulin," Part 2, 753.
31. Chappell to Ducas and Stegen, July 3, 1952, S 14: Polio, Box 12, MPR, MDA.
32. Elmer Bertelsen, "Giving Polio Shots Starts Here," *Houston Chronicle*, July 2, 1952, 1.
33. "Medicine: Betting on G.G.," *Time*, July 14, 1952, http://www.time.com/time/magazine/article/0,9171,822334,00.html, last viewed December 10, 2010.
34. Chappell to Ducas and Stegen, July 3, 1952, S 14: Polio, Box 12, MPR, MDA.
35. Chappell to Ducas and Stegen, July 5, 1952, ibid.
36. "Medicine: Betting on G.G.," *Time*.
37. Ibid.
38. Carleen R. Laurentz, e-mail message to author, June 28, 2010.
39. Chappell to Ducas and Stegen, July 3, 1952, S 14: Polio, Box 12, MPR, MDA.
40. Bertelsen, "Clinics Are Crowded by Children," 1.
41. Chappell to Ducas and Stegen, July 3, 1952, S 14: Polio, Box 12, MPR, MDA.
42. Walter Sneader, *Drug Discovery: A History* (Chichester: Wiley-Blackwell, 2005), 156; R. von den Velden, "Die intrakardiale Injektion," *Münchener medizinische Wochenschrift* 66 (1919): 274–275; G. Kneier, "Uber initiale Adrenalininjektion bei acuter Herzlaehmug," *Deutsche Medizinische Wochenschrift* 47 (1921): 1490–1491.
43. Hammon et al., "Evaluation of Red Cross Gamma Globulin," Part 2, 753.
44. E. R. Shemin, "Anaphylactic Reaction to Gamma-Globulin," *JAMA* 203, 1 (1968): 59.
45. Chappell to Ducas and Stegen, July 3, 1952, S 14: Polio, Box 12, MPR, MDA.
46. Susan L. Smith, *Sick and Tired of Being Sick and Tired: Black Women's Health Activism in America, 1890–1950* (Philadelphia: University of Pennsylvania Press, 1995); Edward H. Beardsley, *A History of Neglect: Health Care for Blacks and Mill Workers in the Twentieth-Century South* (Knoxville: University of Tennessee Press, 1987); Spencie Love, *One Blood* (Chapel Hill: University of North Carolina Press, 1996); Vanessa Northington Gamble, *Making a Place for Ourselves: The Black Hospital Movement* (New York: Oxford University Press, 1995).
47. Pierce and Hammon, "Digest of Duties and Functions," June 6, 1952, S 3: GG FT, Box 3, SSR, MDA, 20.
48. "To Aid Paralysis Victims," *New York Times*, December 25, 1938, 16.

49. Pierce and Hammon, "Digest of Duties and Functions," June 6, 1952, S 3: GG FT, Box 3, SSR, MDA, 20.
50. Stephen E. Mawdsley, "'Dancing on Eggs': Charles H. Bynum, Racial Politics, and the National Foundation for Infantile Paralysis, 1938–1954," *Bulletin of the History of Medicine* 84, 2 (Summer 2010): 217–247.
51. "Charles H. Bynum," [Condensed] Resume, undated [circa 1970s], Dianne H. McDonald private collection, New York.
52. Pierce and Hammon, "Digest of Duties and Functions," June 6, 1952, S 3: GG FT, Box 3, SSR, MDA, 20.
53. "Baytown Man Dies of Polio," *Houston Chronicle*, July 7, 1952, 1, 11.
54. Hammon et al., "Evaluation of Red Cross Gamma Globulin," Part 2, 753.
55. Wooten, *The Polio Years in Texas*, 142.
56. "Clinics Where Polio Shots Can Be Taken," *Houston Chronicle*, July 2, 1952, 10.
57. "Workers Badly Needed," *Houston Informer*, July 5, 1952, 1.
58. Ibid.
59. Susan M. Reverby, ed., *Tuskegee's Truths: Rethinking the Tuskegee Syphilis Study* (Chapel Hill: University of North Carolina Press, 2000); James H. Jones, *Bad Blood: The Tuskegee Syphilis Experiment* (New York: The Free Press, 1981).
60. Becker to Weaver, July 14, 1952, S 14: Polio, Box 12, MPR, MDA.
61. "Demand Grows for Antipolio Test Shots," *Houston Chronicle*, July 2, 1952, 8.
62. Ibid.
63. Becker to Weaver, July 14, 1952, S 14: Polio, Box 12, MPR, MDA.
64. "Medicine: Betting on G.G.," *Time*.
65. "Polio, Measles Shots Same, Doctor Declares," *Houston Chronicle*, July 4, 1952.
66. Stegen to Department Heads, July 7, 1952, S 14: Polio, Box 12, MPR, MDA.
67. Chappell to Ducas and Stegen, July 5, 1952, S 14: ibid.
68. "Demand Grows for Antipolio Test Shots," *Houston Chronicle*, 8.
69. Hammon et al., "Evaluation of Red Cross Gamma Globulin," Part 2, 753.
70. Ibid.
71. "Boy Believed Polio Fatality," *Houston Chronicle*, July 5, 1952, 1.
72. Ibid.
73. "Polio Hits 24," *Houston Chronicle*, July 9, 1952, 1, 10.
74. "Polio Field Study Here Is Extended," *Houston Chronicle*, July 11, 1952, 1, 2.
75. Marshall Verniaud, "Polio Tests Pass Shot Phase Here," *Houston Chronicle*, July 13, 1952, 1, A12.
76. Hammon et al., "Evaluation of Red Cross Gamma Globulin," Part 2, 752–753.
77. Ibid., 754.
78. "Big Response to Program of Inoculation," *Cedar Rapids Gazette*, July 22, 1952, 2.
79. "Sioux City to Be Site of Polio Serum Testing," *Oelwein Daily Register*, July 18, 1952, 1.
80. Joanne Fox, "Caregivers Recall Summer of '52 Polio Epidemic," *Sioux City Journal*, August 18, 2002.
81. Arlene Moltsau, letter to author, June 1, 2010.
82. "Big Response to Program of Inoculation," *Cedar Rapids Gazette*, 2.
83. Fox, "Caregivers Recall Summer of '52."
84. Census of Population, 1950, Characteristics of the Population: Iowa, http://www2.census.gov/prod2/, last viewed January 3, 2015.
85. Hammon et al., "Evaluation of Red Cross Gamma Globulin," Part 2, 754.
86. Ibid.

87. Becker to Weaver, July 14, 1952, S 14: Polio, Box 12, MPR, MDA.
88. Hammon et al., "Evaluation of Red Cross Gamma Globulin," Part 2, 755.
89. Bob Dodsley, "To Inoculate 16,500 Children with Globulin," *Sioux City Journal*, July 18, 1952, 2.
90. "The Polio Tests—Editorial," *Sioux City Journal*, July 20, 1952, 4.
91. "Supervising Nurse Arrives," *Sioux City Journal*, July 19, 1952, 2; "Polio Toll Mounts as Plans Jell for Giant Test Project," *Sioux City Journal*, July 19, 1952, 1.
92. Pierce and Hammon, "Digest of Duties and Functions," June 6, 1952, S 3: GG FT, Box 3, SSR, MDA, 12.
93. "Polio Toll Mounts as Plans Jell for Giant Test Project," *Sioux City Journal*, 1.
94. Dodsley, "To Inoculate 16,500 Children," 1.
95. Hammon et al., "Evaluation of Red Cross Gamma Globulin," Part 2, 755.
96. Weaver to Stegen, August 26, 1952, S 14: Polio, Box 12, MPR, MDA.
97. "Big Response to Program of Inoculation," *Cedar Rapids Gazette*, 2.
98. "Polio Tests Planned at Sioux City," *Cedar Rapids Gazette*, July 18, 1952, 1.
99. "Youngsters Crowd S.C. Polio Clinic," *Council Bluffs Iowa Nonpareil*, July 21, 1952, 1.
100. "Big Response to Program of Inoculation," *Cedar Rapids Gazette*, 2.
101. "Operation Lollipop," *Sioux City Journal*, July 27, 1952, C4.
102. Carol Andersen Clemens, e-mail message to author, May 3, 2010.
103. Ibid.
104. Steve C. Miller, e-mail message to author, August 1, 2010.
105. Hammon et al., "Evaluation of Red Cross Gamma Globulin," Part 2, 757.
106. "Big Response to Program of Inoculation," *Cedar Rapids Gazette*, 2.
107. "End First Phase of Paralysis Test," *Council Bluffs Iowa Nonpareil*, July 27, 1952, 6.
108. Hammon et al., "Evaluation of Red Cross Gamma Globulin," Part 2, 756.
109. "Statement of the NFIP on GG and Vaccine," March 1953, S 14: Polio, Box 13, MPR, MDA, 2.
110. "Gamma Globulin Helps Some against Polio; But Isn't the Answer," *Wall Street Journal*, October 23, 1952, 18; Panel on Allocation of Gamma Globulin, March 5, 1953, S 14: Polio, Box 13, MPR, MDA.
111. W. M. Hammon, L. L. Coriell, and J. Stokes Jr., "Evaluation of Red Cross Gamma Globulin as a Prophylactic Agent for Poliomyelitis: 1. Plan of Controlled Field Tests and Results of the 1951 Pilot Study in Utah," *JAMA* 150 (1952): 739.
112. Ibid.
113. Ibid., 744.
114. Susan E. Lederer and Jonathan D. Moreno, "Revising the History of Cold War Research Ethics," *Kennedy Institute of Ethics Journal* 6, 3 (September 1996): 223–237; David Pacchioli, "Subjected to Science," *Research/Penn State* 17, 1 (March 1996), http://www.rps.psu.edu/mar96/science.html, last viewed March 10, 2011.
115. Hammon et al., "Evaluation of Red Cross Gamma Globulin," Part 1, 742, 743, 748. See also W. M. Hammon, L. L. Coriell, P. F. Wehrle, and J. Stokes Jr., "Evaluation of Red Cross Gamma Globulin as a Prophylactic Agent for Poliomyelitis: 4. Final Report of Results Based on Clinical Diagnoses," *JAMA* 151 (1953): 1272–1285.
116. Trevor Pinch, "'Testing—One, Two, Three . . . Testing!': Toward a Sociology of Testing," *Science, Technology, & Human Values* 18 (Winter 1993): 29–30.
117. Hammon et al., "Evaluation of Red Cross Gamma Globulin," Part 1, 743.
118. An-Wen Chan, Asbørn Hróbjartsson, Mette T. Haahr, Peter C. Gøtzsche, Douglas G. Altman, "Empirical Evidence for Selective Reporting of Outcomes in Randomized

Trials: Comparison of Protocols to Published Articles," *JAMA* 291, 20 (2004): 2457–2465.

119. Hammon et al., "Evaluation of Red Cross Gamma Globulin," Part 1, 742.
120. Rivers in Saul Benison, *Tom Rivers: Reflections on a Life in Medicine and Science* (Cambridge, Mass.: MIT Press, 1967), 484.
121. Ibid., 484–485.
122. Ibid., 486–487.
123. Marguerite Clark, "How They're Closing In on Polio," *Popular Science*, May 1953, 140.
124. "Gamma Globulin to the Rescue," *Life* magazine, November 3, 1952, 52.
125. "Tests Show Injections Protect Children from Paralytic Polio," *El Paso Herald-Post*, October 22, 1952, 1.
126. Richard Carter, *Breakthrough: The Saga of Jonas Salk* (New York: Pocket Books, 1967), 102.
127. Zimmerer to Hammon, November 25, 1952, S 3: GG FT, Box 3, SSR, MDA.
128. Rhoda Truax, *True Adventures of Doctors* (Boston: Little, Brown, 1954), 207.
129. NFIP News Release to City Editors, November 11, 1952, S 14: Polio, Box 12, MPR, MDA.
130. Clark, "Closing In on Polio," 137.
131. Guy Oakes, *The Imaginary War: Civil Defense and Cold War Culture* (Oxford: Oxford University Press, 1994); Richard Severo, "Ed Herlihy, 89, a Voice of Cheer and Cheese," February 2, 1999, *New York Times*, http://www.nytimes.com/1999/02/02/arts/ed-herlihy-89-a-voice-of-cheer-and-cheese.html, last viewed April 2, 2010.
132. *Operation Marbles and Lollipops*, NFIP, DVD (1952; New York: MDA, 2009).
133. Ibid.
134. Susan Lederer, "Hollywood and Human Experimentation," in *Medicine's Moving Pictures*, ed. Leslie J. Reagan, Nancy Tomes, and Paula A. Treichler (Rochester, N.Y.: University of Rochester Press, 2007), 285–286.
135. Susan E. Lederer, *Subjected to Science: Human Experimentation in America before the Second World War* (Baltimore: Johns Hopkins University Press, 1997).
136. *Operation Marbles and Lollipops*, MDA.
137. Ibid.
138. Lederer, "Hollywood and Human Experimentation," in Reagan et al., *Medicine's Moving Pictures*, 287.
139. *Operation Marbles and Lollipops*, MDA.
140. Naomi Rogers, "Race and the Politics of Polio: Warm Springs, Tuskegee, and the March of Dimes," *AJPH* 97 (2007): 784–795.
141. Cramer to Hammon, January 15, 1953, St65p, Hammon, W. McD. #30, Joseph Stokes, Jr., Papers, APS.
142. Mawdsley, "'Dancing on Eggs,'" 242.
143. Michael L. Krenn, ed., *Race and U.S. Foreign Policy during the Cold War* (New York: Garland Publishing, 1998).

Chapter 6 — The National Experiment

1. "140 Campers Get Polio Immunizer," *New York Times*, August 23, 1953, 49.
2. Brenda Serotte, *The Fortune Teller's Kiss* (Lincoln: University of Nebraska Press, 2006), 76.
3. Johannes Ipsen, "Administrative Problems in the Use of Poliomyelitis Immune Globulin," *AJPH* 43 (September 1953): 1101–1110.

4. Zuoyue Wang, *In Sputnik's Shadow: The President's Science Advisory Committee and Cold War America* (New Brunswick, N.J.: Rutgers University Press, 2008), 34–36.
5. Burton I. Kaufman, *The Korean War: Challenges in Crisis, Credibility, and Command* (New York: McGraw-Hill, 1997).
6. General Statement of the American Red Cross Regarding the Procurement of Gamma Globulin, January 7, 1953, S 14: Polio, Box 12, Medical Program Records (henceforth MPR), March of Dimes Archives (henceforth MDA).
7. "Gamma Globulin," *American National Red Cross 1953 Annual Report*, American Red Cross Archives, Washington, D.C., 22.
8. Winternitz to Cummings, January 26, 1953, S 14: Polio, Box 13, MPR, MDA; General Statement of NFIP Regarding Plan for Procurement and Use of Gamma Globulin during First Six Months of 1953, January 7, 1953, S 14: Polio, Box 12, MPR, MDA, 2.
9. "Washington Report on the Medical Sciences," January 12, 1953, S 14: Polio, Box 12, MPR, MDA; Stephanie to Barrows, February 20, 1953, S 14: ibid.; "Meeting of Lafayette Building," March 9, 1953, Box 1215, National Archives and Records Administration, College Park, Maryland (henceforth NARA).
10. NFIP Stand on GG, September 30, 1953, S 14: Polio, Box 13, MPR, MDA.
11. General Statement of the American Red Cross, January 7, 1953, S 14: Polio, Box 12, MPR, MDA, 2.
12. U.S. Public Health Service, "Panel on Allocation of Gamma Globulin," January 5, 1953, S 14: Polio, Box 13, MPR, MDA.
13. News Release, December 11, 1953, S 14: ibid.
14. "Maryland State Department of Health: Plan for Gamma Globulin Distribution," April 17, 1953, S 14: ibid.
15. Hilleboe to O'Connor, November 19, 1953, S 3: GG FT, Box 4, Surveys and Studies Records (henceforth SSR), MDA.
16. Barrows to O'Connor, March 23, 1953, S 14: Polio, Box 13, MPR, MDA; "Gamma Globulin Activity, 1953–1958," 1958, SSPVR, Box 1, S3: Gamma Globulin Activity (henceforth GGA), MDA.
17. Winternitz to Cummings, January 26, 1953, S 14: Polio, Box 13, MPR, MDA; Recommendations of the Panel on Allocation of Gamma Globulin, January 20, 1953, S 14: ibid., 2.
18. U.S. Public Health Service, "Panel on Allocation," January 5, 1953, S 14: ibid.
19. Ibid.
20. Kumm to NFIP, March 5, 1953, S 14: Polio, Box 13, MPR, MDA.
21. Ibid.
22. McGinnes to Barrows, January 29, 1953, S 14: Polio, Box 12, MPR, MDA.
23. "Evaluation of Gamma Globulin in the Prophylaxis of Paralytic Polio in the United States During 1953," S 3: GG FT, Box 4, SSR, MDA, ii.
24. "Meetings of Advisory Committee on a National Program for Evaluation of Gamma Globulin in the Prophylaxis of Poliomyelitis," May 30, 1953, GG Program (1953), S 14: Polio, Box 13, MPR, MDA, 2.
25. Rivers in Saul Benison, *Tom Rivers: Reflections on a Life in Medicine and Science* (Cambridge, Mass.: MIT Press, 1967), 485–486.
26. Ibid.
27. News Release, April 20, 1953, S 14: Polio, Box 13, MPR, MDA.
28. "Polio Group Assails Handling of Globulin," *Washington Post*, April 20, 1953, 5.
29. "Draft Foundation Message," April 1953, S 14: Polio, Box 12, MPR, MDA, 2.

30. Earl Ubell, "Upstate Polio Shots Costliest Ever," *New York Herald Tribune*, July 10, 1953.
31. News Release, August 26, 1953, S 14: Polio, Box 13, MPR, MDA.
32. News Release, February 1953, S 14: Polio, Box 12, MPR, MDA.
33. News Release, February 16, 1953, S 14: ibid.
34. Phone Calls, January 12, 1953, S 14: ibid.
35. Crawford to Stegen, February 16, 1953, S 14: ibid.
36. Culp to Reese, April 23, 1953, S 14: Polio, Box 13, MPR, MDA.
37. Ellen Schrecker, *The Age of McCarthyism* (New York: Palgrave Macmillan, 2002); Ted Morgan, *Reds: McCarthyism in Twentieth-Century America* (New York: Random House, 2004).
38. Revised Recommendations of the Panel on Allocation of Gamma Globulin, February 14, 1953, S 14: Polio, Box 12, MPR, MDA, 3.
39. Van Riper to Chapter Chairmen, May 5, 1953, S 14: Polio, Box 13, MPR, MDA.
40. Barrows to Davis, February 25, 1953, S 14: Polio, Box 12, MPR, MDA.
41. Armstrong to Grant, March 2, 1953, Box 1214, Gamma Globulin, NARA.
42. Martin Halliwell, *Therapeutic Revolutions: Medicine, Psychiatry, and American Culture, 1945–1970* (New Brunswick, N.J.: Rutgers University Press, 2014); Jonathan Engel, *Doctors and Reformers: Discussion and Debate over Health Policy, 1925–1950* (Columbia: University of South Carolina Press, 2002), chapter 8.
43. Ducas to Barrows, January 30, 1953, S 14: Polio, Box 12, MPR, MDA.
44. Ibid.
45. London to Barrows, April 2, 1953, S 14: Polio, Box 13, MPR, MDA.
46. "$7,000,000 Goal Set by Red Cross Here: Fund Drive to Start March 1," *New York Times*, January 22, 1953, 25.
47. Voss to O'Connor, January 23, 1953, S 14: Polio, Box 12, MPR, MDA.
48. Ibid.
49. Voss to Chapter Chairmen, January 28, 1953, S 14: ibid.
50. "Red Cross Seeks 7 Million in Drive Here," *New York Times*, February 17, 1953, 24.
51. "March Proclaimed Red Cross Month," *New York Times*, February 28, 1953, 14.
52. Anthony Leviero, "Eisenhower Pushes Drive by Red Cross," *New York Times*, January 30, 1953, 27; "Eisenhower Asks Gifts to Red Cross," *New York Times*, February 26, 1953, 22.
53. "First Lady Spurs Red Cross Drives," *New York Times*, March 22, 1953, 2.
54. News Release, January 8, 1953, S 14: Polio, Box 12, MPR, MDA.
55. News Release, January 1, 1953, S 14: Polio, Box 13, MPR, MDA.
56. Bruce Cummings, ed., *Child of Conflict: The Korean-American Relationship, 1943–1953* (Seattle: University of Washington Press, 1983), and *The Origins of the Korean War* (Princeton: Princeton University Press, 1981).
57. News Release, January 1, 1953, S 14: Polio, Box 13, MPR, MDA.
58. Ducas to Barrows, January 30, 1953, S 14: Polio, Box 12, MPR, MDA; Ipsen, "Administrative Problems," 1101–1110.
59. Dick Preston, "Polio Victims' Kin to Get Rare Serum," *Scripps-Howard News Service*, March 12, 1953, GG Program (1953), S 14: Polio, Box 13, MPR, MDA.
60. Ducas to Publishers, February 24, 1953, S 14: Polio, Box 12, MPR, MDA.
61. "Polio Message: Gamma Globulin—1953," S 14: Polio, Box 13, MPR, MDA.
62. David M. Oshinsky, *Polio: An American Story: The Crusade That Mobilized the Nation against the 20th Century's Most Feared Disease* (Oxford: Oxford University Press, 2005), chapter 7.

63. Statement of the NFIP on GG and Vaccine, March 1953, S 14: Polio, Box 13, MPR, MDA, 4.
64. Leonard Engel, "Polio: New Weapons and New Hope," *New York Times*, May 31, 1953, SM11.
65. News Release, March 9, 1953, S 14: Polio, Box 13, MPR, MDA.
66. Ipsen, "Administrative Problems," 1101–1110.
67. Ibid.
68. Stegen to Holland, September 3, 1953, S 3: GG FT, Box 3, SSR, MDA.
69. Ducas to Barrows, January 30, 1953, S 14: Polio, Box 12, MPR, MDA.
70. Ibid.
71. "NFIP Public Relations Procedures in Mass GG Inoculations," July 28, 1953, S 14: Polio, Box 13, MPR, MDA.
72. NFIP to Voss, January 1953, S 14: ibid.
73. "NFIP Public Relations Procedures," July 28, 1953, S 14: ibid.
74. Ibid., 5.
75. "Evaluation of Gamma Globulin," S 3: GG FT, Box 4, SSR, MDA, 71.
76. "Gamma Globulin Season," *Time*, July 13, 1953; James Liston, "How Will We Fight Polio This Year?" *Better Homes & Gardens*, June 1953.
77. "Victim Was Stricken at Youth Camp," *Appeal-Democrat (California)*, July 1, 1953.
78. "Medicine: Capsules," *Time*, May 18, 1953.
79. "Offer of Bribe for Anti-Polio 'Shot' Is Bared," *Chicago Tribune*, August 19, 1953, 21.
80. John Geiger, "Gamma Globulin Checks Polio Epidemic," *Texas Avalanche*, July 13, 1953, 10.
81. D. G. Gill, "Gamma Globulin in a Poliomyelitis Outbreak in Montgomery, Alabama, 1953," *Public Health Reports* 68, 11 (November 1953): 1021–1024.
82. Dick Preston, "U.S. Blood Donors Share in Mass Fight on Polio," *Memphis Press Scimitar*, June 3, 1953.
83. Gill, "Gamma Globulin in a Poliomyelitis Outbreak," 1021–1024.
84. Preston, "U.S. Blood Donors Share in Mass Fight on Polio."
85. Waldemar Kaempffert, "Science in Review," *New York Times*, July 12, 1953, E9.
86. "The Polio Fight in Alabama," *New York Herald Tribune*, July 2, 1953, 14.
87. Preston, "U.S. Blood Donors Share in Mass Fight on Polio."
88. Charles M. Cameron, "Organizing Mass Gamma Globulin Clinics in Three North Carolina Counties," *Public Health Reports* 68, 11 (November 1953): 1025–1033.
89. Culp to Barrows, July 13, 1953, S 3: GG FT, Box 3, SSR, MDA.
90. Earl Ubell, "Upstate Polio Shots Costliest Ever," *New York Herald Tribune*, July 10, 1953.
91. Ibid.
92. U.S. Public Health Service, "Panel on Allocation," January 5, 1953, S 14: Polio, Box 13, MPR, MDA.
93. Bulletin, Panel on Allocation of GG, March 1953, S 14: ibid.
94. Kumm to Weaver, August 14, 1953, S 3: GG FT, Box 3, SSR, MDA, 3.
95. Ducas to Barrows, October 1, 1953, S 14: Polio, Box 13, MPR, MDA; "Polio Preventatives Held Nonexistent," *New York Times*, September 20, 1993, 28.
96. Elizabeth W. Etheridge, *Sentinel for Health: A History of the Centers for Disease Control* (Berkeley: University of California Press, 1992), 70.
97. Kumm to Barrows, February 25, 1954, S 3: GG FT, Box 4, SSR, MDA.
98. "Members of National Advisory Committee for the Evaluation of Gamma Globulin," January 20, 1954, S 3: ibid.

99. Chief, Epidemic Intelligence Officers Unit, September 8, 1953, S 3: GG FT, Box 3, SSR, MDA.
100. "Meeting of Medical Practitioner Committee on Gamma Globulin Distribution," DHEW, September 10, 1953, S 3: ibid., 5.
101. Langmuir to Kumm, December 22, 1953, S 3: GG FT, Box 4, SSR, MDA.
102. Crawford to Ducas, January 7, 1954, S 3: ibid.
103. Ibid.
104. Barrows to O'Connor, January 21, 1954, S 14: Polio, Box 13, MPR, MDA.
105. Marcia Lynn Meldrum, "Departures from the Design: The Randomized Clinical Trial in Historical Context, 1946–1970" (PhD diss., State University of New York, 1994), 92–93.
106. Glasser to O'Connor, January 19, 1954, S 14: Polio, Box 13, MPR, MDA.
107. Barrows to O'Connor, January 21, 1954, S 14: ibid.
108. Ducas to O'Connor, January 8, 1954, S 14: ibid.
109. National Advisory Committee for the Evaluation of Gamma Globulin in the Prophylaxis of Poliomyelitis, *An Evaluation of the Efficacy of Gamma Globulin in the Prophylaxis of Paralytic Poliomyelitis as Used in the United States, 1953*, Public Health Monograph No. 20 (Washington, D.C.: US GPO, 1954) (PHS publication no. 358).
110. Memorandum, Members of National Advisory Committee for the Evaluation of Gamma Globulin, January 20, 1954, S 3: GG FT, Box 4, SSR, MDA.
111. DHEW News Release, February 23, 1954, S 3: ibid.
112. Memorandum, Members of National Advisory Committee for the Evaluation of Gamma Globulin, January 20, 1954, S 3: ibid.
113. Ibid.
114. "Medicine: Decision Reversed," *Time*, March 1, 1954, http://www.time.com/time/magazine/article/0,9171,819505,00.html, last viewed January 10, 2011.
115. Ibid.; "Find Gamma Globulin No Help in Polio," *Chicago Tribune*, February 23, 1954, 1.
116. "Gamma Globulin of Doubtful Value for Polio Prophylaxis," *North Carolina Medical Journal* 15 (1954): 137, 154.
117. Ducas and St. John to Regional Representatives, October 1, 1953, S 14: Polio, Box 13, MPR, MDA.
118. Ibid.
119. Young to Ducas, October 19, 1953, October 1, 1953, S 14: ibid.
120. Richard Carter, *Breakthrough: The Saga of Jonas Salk* (New York: Trident Press, 1966), 102–103.
121. NFIP News Release, February 23, 1954, S 14: Polio, Box 13, MPR, MDA.
122. Editorial, "The Polio Season," *Life* magazine, May 24, 1954, 28.
123. NFIP News Release, June 10, 1954, S 14: Polio, Box 13, MPR, MDA.
124. "Longines-Wittnauer with Basil O'Connor," 95906, LW-LW-325, 1954, v.8, NARA.
125. Ibid.
126. Ibid.
127. Hammon to Smadel, May 8, 1953, Meetings, Army Epidemiological Board, Hauck Center for the Albert B. Sabin Archives, University of Cincinnati Libraries, Cincinnati, Ohio (henceforth HCASA).
128. William McD. Hammon et al., "Comparative Studies on Patterns of Family Infections with Polioviruses and ECHO Virus Type I on an American Military Base in the Philippines," *AJPH* 47 (July 1957): 802–811.
129. Hammon to Sabin, May 11, 1953, Meetings, Army Epidemiological Board, HCASA; Hammon to Sabin, December 3, 1956, Meetings, HCASA.

130. Ducas to O'Connor, June 16, 1954, S 14: Polio, Box 13, MPR, MDA.
131. Stegen to Ducas, June 14, 1954, S 14: ibid.
132. Hammon, Press Release, February 24, 1954, S 14: ibid.
133. Ibid.
134. Stokes to Kumm, March 1, 1954, S 3: GG FT, Box 4, SSR, MDA. W. M. Hammon, L. L. Coriell, E. H. Ludwig, R. M. McAllister, A. E. Greene, G. E. Sather, and P. F. Wehrle, "Evaluation of Red Cross Gamma Globulin as a Prophylactic Agent for Poliomyelitis. 5. Reanalysis of Results Based on Laboratory-Confirmed Cases," *JAMA* 156, 1 (1954): 21–27.
135. NFIP Memorandum for Basil O'Connor, August 1954, S 3: GG FT, Box 4, SSR, MDA.
136. Lutey to Barrie, November 16, 1953, S 14: Polio, Box 13, MPR, MDA.
137. Kenneth S. Landauer, "Report to Physicians," Summer 1954, S 3: GG FT, Box 4, SSR, MDA.
138. Oshinsky, *Polio: An American Story*, 157–159.
139. Christopher J. Rutty, "Do Something! . . . Do Anything!: Poliomyelitis in Canada, 1927–1962" (PhD diss., University of Toronto, 1995); Oshinsky, *Polio: An American Story*, 191–192.
140. Kumm to NFIP, January 30, 1954, S 14: Polio, Box 13, MPR, MDA; Glasser to Bell, September 1953, S 14: ibid.
141. Howard A. Rusk, "Confusion Over Programs to Combat Polio Justified," *New York Times*, March 14, 1954, 64.
142. Kumm to Barrows, February 25, 1954, S 3: GG FT, Box 4, SSR, MDA.
143. Barrows to O'Connor, February 19, 1954, S 14: Polio, Box 13, MPR, MDA.
144. "Key West Fights Polio," *New York Times*, May 3, 1954, 12; "Polio Epidemic Fought in West," *New York Times*, July 10, 1954, 15.
145. Ducas to O'Connor, June 29, 1954, S 14: Polio, Box 13, MPR, MDA.
146. Osborne to Ducas, June 7, 1954, S 14: ibid.
147. Barrows to O'Connor, June 23, 1954, S 14: ibid.
148. Ducas to Wrigley, December 14, 1953, S 14: ibid.
149. "When Help Was Needed, Your Dollars Were There," *New York Times*, December, 27, 1954, 11.
150. Meldrum, "Departures from the Design," 129; John Troan, *Passport to Adventure: Or, How a Typewriter from Santa Led to an Exciting Lifetime Journey* (Pittsburgh: Cold-Comp, 2000), 213; Paul A. Offit, *The Cutter Incident: How America's First Polio Vaccine Led to the Growing Vaccine Crisis* (New Haven: Yale University Press, 2005), 36, 38.
151. Meldrum, "Departures from the Design," 129; Troan, *Passport to Adventure*, 213; Offit, *The Cutter Incident*, 36, 38.
152. Meldrum, "Departures from the Design," 93.
153. Crawford to Ducas, January 7, 1954, S 3: GG FT, Box 4, SSR, MDA.
154. Oshinsky, *Polio: An American Story*, chapter 12.
155. Ibid., 191.
156. John R. Paul, *A History of Poliomyelitis* (New Haven: Yale University Press, 1971), 426–427; Carter, *Breakthrough*, 238; Jane S. Smith, *Patenting the Sun: Polio and the Salk Vaccine* (New York: William Morrow, 1990), 228.
157. Howard A. Rusk, "Confusion Over Programs," 64.
158. NFIP Polio Precautions, No. 31, April 1954, S 14: Polio, Box 13, MPR, MDA.
159. Van Riper to NFIP Staff, September 8, 1954, S 14: ibid.; Gorrell to Van Riper, August 24, 1954, S 14: ibid.

160. Barrows to NFIP Staff, September 30, 1954, S 14: ibid.; Francis to Physicians, August 6, 1954, S 14: ibid.
161. Glasser to Bell, September 1953, S 14: ibid.
162. "Polio Test Report Set For April 12," *New York Times*, March 23, 1955, 33; William L. Laurence, "Salk Polio Vaccine Proves Success; Millions Will Be Immunized Soon; City Schools Begin Shots April 25," *New York Times*, April 13, 1955, 1; Jeffrey Kluger, *Splendid Solution: Jonas Salk and the Conquest of Polio* (New York: G. P. Putnam's Sons, 2004), 294; Paul, *A History of Poliomyelitis*, 432.
163. Leigh Flower Bonner, Houston, e-mail message to author, August 5, 2010.
164. Smith, *Patenting the Sun*, 386.
165. "One-Shot Vaccine for Measles," *Time*, February 19, 1965, http://www.time.com/time/magazine/article/0,9171,940965,00.html, last viewed July 1, 2011.
166. "Infectious Diseases: German Measles Epidemic," *Time*, April 24, 1964, http://www.time.com/time/magazine/article/0,9171,870897,00.html, last viewed July 2, 2011.
167. S. Iwarson and K. Stenqvist, "Tourist Hepatitis and Gamma Globulin Prophylaxis," *Scandinavian Journal of Infectious Diseases* 8, 3 (1976): 143–145.
168. "Medicine: Transfusion for Hepatitis," *Time*, November 8, 1968, http://www.time.com/time/magazine/article/0,9171,902513,00.html, last viewed July 8, 2011.
169. "Travel Advisory; Gamma Globulin In Short Supply," *New York Times*, March 5, 1995, http://www.nytimes.com/1995/03/05/travel/travel-advisory-gamma-globulin-in-short-supply.html, last viewed July 5, 2011.
170. Paul A. Offit, *Vaccinated: One Man's Quest to Defeat the World's Deadliest Diseases* (Washington, D.C.: Smithsonian, 2007), 107.
171. Maurice B. Strauss, ed., *Familiar Medical Quotations* (Little, Brown, 1968), 625.

Bibliography

Archives

American Philosophical Society (APS), Philadelphia, Pennsylvania
American Red Cross Archives (ARC), Washington, D.C.
Bancroft Library, University of California, Berkeley, California
Bentley Historical Library (BHL), Ann Arbor, Michigan
Claude Moore Health Sciences Library, University of Virginia, Charlottesville, Virginia
Columbia University Archives, New York
Donald C. Harrison Health Sciences Library, University of Cincinnati, Ohio
Hauck Center for the Albert B. Sabin Archives (HCASA), University of Cincinnati Libraries, Cincinnati, Ohio
James Lind Library, The Royal College of Physicians of Edinburgh
Mandeville Special Collections Library (MSCL), University of California, La Jolla, California
March of Dimes Archives (MDA), White Plains, New York
National Archives and Records Administration (NARA), College Park, Maryland
National Library of Medicine, Bethesda, Massachusetts
University of Minnesota Archives, Minneapolis, Minnesota
Yale University Library, New Haven, Connecticut

Filmography

Dime Power. NFIP. DVD. 1954; MDA, 2007.
Interim Report. NFIP. DVD. 1956; MDA, 2008.
Johnny—A Filmstrip. NFIP. DVD. 1956; MDA, 2008.
Longines-Wittnauer with Basil O'Connor. National Archives and Records Administration. ARC Identifier 95906 / Local Identifier LW-LW-325–1954, v.8.
Operation Marbles and Lollipops. NFIP. DVD. 1952; MDA, 2009.
Mother's March on Polio. NFIP. DVD. 1951; MDA, 2006.
Negro Newsreel. NFIP. DVD. 1956; MDA, 2008.
Newsreel. NFIP. DVD. 1957; MDA, 2008.
Offit, Paul A., MD. *The Cutter Incident: Lessons from the Past*. DVD. Collection from A Scientific Symposium Commemorating the 50th Anniversary of the Development of the Polio Vaccine. Pittsburgh, Pa.: University of Pittsburgh, 2005.
Polio and the Vaccine. NFIP. DVD. 1956; MDA, 2008.
Salk, Peter L., MD. *Memoirs of My Father: Personal Reflections on Jonas Salk*. DVD. Collection from A Scientific Symposium Commemorating the 50th Anniversary of the Development of the Polio Vaccine. Pittsburgh, Pa.: University of Pittsburgh, 2005.
Unconditional Surrender. NFIP. DVD. 1956; MDA, 2008.

Articles, Dissertations, and Monographs

Aitken, Hugh G. J. *Scientific Management in Action: Taylorism at Watertown Arsenal, 1908–1915*. Princeton: Princeton University Press, 1985.
Aitken, Sally, Helen D'Orazio, and Stewart Valin, eds. *Walking Fingers: The Story of Polio and Those Who Lived with It*. Montreal: Vehicule Press, 2004.

Albarelli, H. P. *A Terrible Mistake: The Murder of Frank Olson and the CIA's Secret Cold War Experiments*. Walterville, Ore.: TrineDay, 2008.

Allen, Arthur. *Vaccine: The Controversial Story of Medicine's Greatest Lifesaver*. New York: W. W. Norton, 2007.

Altman, Lawrence K. *Who Goes First? The Story of Self-Experimentation in Medicine*. Berkeley: University of California Press, 1998.

American Heritage Dictionary of the English Language. 4th ed. Boston: Houghton Mifflin Company, 2004.

American National Red Cross, *The American Red Cross Blood Donor Service during World War II, Its Organization and Operation*. Washington, D.C.: American National Red Cross, 1946.

Anderson, Ann. *Snake Oil, Hustlers, and Hambones: The American Medicine Show*. Jefferson, N.C.: McFarland and Co., 2000.

Andrews, F. Emerson. *Philanthropic Giving*. Philadelphia: Wm. F. Fell Co. Printers, 1950.

Apple, Rima M. *Perfect Motherhood: Science and Childrearing in America*. New Brunswick, N.J.: Rutgers University Press, 2006.

Armstrong, Charles, C. H. Best, A. Baird Hastings, Carl Caskey Speidel, Howard B. Lewis, Walter J. Meek, and Rebecca C. Lancefield. *The Harvey Lectures: 1940–1941*. Series 36. Lancaster, Pa.: Science Press Printing Co., 1941.

Arnold, David. *Colonizing the Body: State Medicine and Epidemic Disease in Nineteenth-Century India*. Berkeley: University of California Press, 1993.

Austin, Joe, and Michael Nevin Willard. *Generations of Youth: Youth Cultures and History in Twentieth-Century America*. New York: New York University Press, 1998.

Bader, George B. "The Intramuscular Injection of Adult Whole Blood as Prophylactic against Measles: With a Report on the Literature." *JAMA* 93, 9 (August 1929): 668–670.

Baghdady, Georgette, and Joanne M. Maddock. "Marching to a Different Mission." *Stanford Social Innovation Review* 6, 2 (Spring 2008): 61–65.

Baker, Robert B., Arthur L. Caplan, Linda L. Emanuel, and Stephen R. Latham, eds. *The American Medical Ethics Revolution: How the AMA's Code of Ethics Has Transformed Physicians' Relationships to Patients, Professionals, and Society*. Baltimore: Johns Hopkins University Press, 1999.

Bannister, Betty. *Trapped: A Polio Victim's Fight for Life*. Saskatoon: Western Producer Prairie Books, 1975.

Barenberg, L. H., J. M. Lewis, and W. H. Messer. "Measles Prophylaxis: Comparative Results with the Use of Adult Blood, Convalescent Serum, and Immune Goat Serum." *JAMA* 95, 1 (1930): 4–8.

Barnes, Barry, David Bloor, and John Henry. *Scientific Knowledge: A Sociological Analysis*. Chicago: University of Chicago Press, 1996.

Barr, Donald A. *Introduction to U.S. Health Policy: The Organization, Financing, and Delivery of Health Care in America*. 2nd ed. Baltimore: Johns Hopkins University Press, 2007.

Barth, Kelly, ed. *Human Medical Trials*. Opposing Viewpoints: At Issue in History Series. Detroit, Mich.: Thomson Gale, 2005.

Barton, Barbara Ann. "The Relationship between Adaptation to Disability, and Sexual and Body Esteem in Women with Polio." PhD diss., Michigan State University, 2005.

Beardsley, Edward H. *A History of Neglect: Health Care for Blacks and Mill Workers in the Twentieth-Century South*. Knoxville: University of Tennessee Press, 1987.

Benison, Saul. "Poliomyelitis and the Rockefeller Institute: Social Effects and Institutional Response." *Journal of the History of Medicine* 29 (1974): 74–92.

———. *Tom Rivers: Reflections on a Life in Medicine and Science.* Cambridge, Mass.: MIT Press, 1967.

Bennett, James T., and Thomas J. Dilorenzo. *Unhealthy Charities: Hazardous to Your Health and Wealth.* New York: HarperCollins, 1994.

Berk, L. B. "Polio Vaccine Trials of 1935." *Transactions & Studies of the College of Physicians of Philadelphia* 11 (1989): 321–326.

Berridge, Virginia, and Kelly Loughlin, eds. *Medicine, the Market, and the Mass Media: Producing Health in the Twentieth Century.* London: Routledge, 2005.

Berry, Donald A. "A Case for Pragmatism in Clinical Trials." *Statistics in Medicine* 12 (1993): 1377–1393.

Black, Allida M. *Casting Her Own Shadow: Eleanor Roosevelt and the Shaping of Postwar Liberalism.* New York: Columbia University Press, 1996.

Black, Conrad. *Franklin Delano Roosevelt: Champion of Freedom.* New York: Public Affairs, 2003.

Black, Kathryn. *In the Shadow of Polio: A Personal and Social History.* Cambridge, Mass.: Perseus Publishing, 1996.

Blumberg, Baruch S. *Hepatitis B: The Hunt for a Killer Virus.* Princeton: Princeton University Press, 2003.

Bookchin, Debbie, and Jim Schumacher. *The Virus and the Vaccine: The True Story of a Cancer-Causing Monkey Virus, Contaminated Polio Vaccine, and the Millions of Americans Exposed.* New York: St. Martin's Press, 2004.

Bourdelais, Patrice. *Epidemics Laid Low: A History of What Happened in Rich Countries.* Translated by Bart K. Holland. Baltimore: Johns Hopkins University Press, 2006.

Boyd, T. E. "Immunization against Poliomyelitis." *Bacteriological Reviews (Supplement)* 17 (1953): 339–448.

Brandt, Allan M. *The Cigarette Century: The Rise, Fall, and Deadly Persistence of the Product that Defined America.* New York: Basic Books, 2007.

———. *No Magic Bullet: A Social History of Venereal Disease in the United States since 1880.* New York: Oxford University Press, 1985, 1987.

———. "Polio, Politics, Publicity, and Duplicity: Ethical Aspects in the Development of the Salk Vaccine." *International Journal of Health Services* 8, 2 (1978): 257–270.

Briggs, Laura. *Reproducing Empire: Race, Sex, Science, and US Imperialism in Puerto Rico.* Berkeley: University of California Press, 2002.

Brown, E. Richard. *Rockefeller Medicine Men: Medicine and Capitalism in America.* Berkeley: University of California Press, 1979, 1980.

Brown, Russell W., and James M. Henderson. "The Mass Production and Distribution of HeLa Cells at Tuskegee Institute, 1953–55." *Journal of the History of Medicine and Allied Sciences* 38, 4 (October 1983): 415–431.

Brownlee, Alexander K. "Statistics of the 1954 Polio Vaccine Trials." *Journal of the American Statistical Association* 50 (1955): 1005–1013.

Brownlee, Shannon. *Overtreated: Why Too Much Medicine Is Making Us Sicker and Poorer.* New York: Bloomsbury USA, 2007.

Bruhl, H. H. "Adverse Reaction to Large Doses of Human Immune Serum Globulin (ISG)." *Minnesota Medicine* 60, 9 (1977): 673–676.

Bruno, Richard L. *The Polio Paradox: What You Need to Know.* New York: Warner Books, 2002.

Brunton, Deborah. *The Politics of Vaccination: Practice and Policy in England, Wales, Ireland, and Scotland, 1800–1874.* Rochester, N.Y.: University of Rochester Press, 2008.

Burke, Donald S. "Lessons Learned from the 1954 Field Trial of Poliomyelitis Vaccine." *Clinical Trials* 1 (2004): 3–5.

Burnham, John C. *How Superstition Won and Science Lost: Popularizing Science and Health in the United States.* New Brunswick, N.J.: Rutgers University Press, 1987.

———. *What Is Medical History?* Cambridge: Polity Press, 2005.

Caplan, Arthur L., James J. McCartney, and Dominic A. Sisti, eds. *Health, Disease, and Illness: Concepts in Medicine.* Washington, D.C.: Georgetown University Press, 2004.

Carney, Scott. *The Red Market: On the Trail of the World's Organ Brokers, Bone Thieves, Blood Farmers, and Child Traffickers.* New York: William Morrow, 2011.

Carrell, Jennifer Lee. *The Speckled Monster: A Historical Tale of Battling Smallpox.* New York: Dutton, 2003.

Carter, Richard. *Breakthrough: The Saga of Jonas Salk.* New York: Pocket Books, 1967.

———. *The Gentle Legions.* New York: Doubleday & Company, 1961.

Castagnoli, William G., Frank Hughes, John Kallir, Michael J. Lyons, and Ron Pantello. *Medicine Ave.: The Story of Medical Advertising in America.* Huntington, N.Y.: Medical Advertising Hall of Fame, 1999.

Chappell, Edith P., and John F. Hume. "A Black Oasis: Tuskegee's Fight against Infantile Paralysis, 1941–1975." New York: March of Dimes Birth Defects Foundation, 1987.

Chaturvedi, Gitanjali. *The Vital Drop: Communication for Polio Eradication in India.* New Delhi: Chaman Enterprises, 2008.

Clark, William R. *At War Within: The Double-Edged Sword of Immunity.* New York: Oxford University Press, 1995.

Clawson, Mary Ann. *Constructing Brotherhood: Class, Gender, and Fraternalism.* Princeton: Princeton University Press, 1989.

Cohn, Victor. *Four Billion Dimes.* Minneapolis, Minn.: Minneapolis Star and Tribune, 1955.

———. *Sister Kenny: The Woman Who Challenged the Doctors.* Minneapolis: University of Minnesota Press, 1975.

Colgrove, James. *State of Immunity: The Politics of Vaccination in Twentieth-Century America.* Berkeley: University of California Press, 2006.

Comacchio, Cynthia, Janet Golden, and George Weisz, eds. *Healing the World's Children: Interdisciplinary Perspectives on Child Health in the Twentieth Century.* Montreal: McGill-Queen's University Press, 2008.

Connelly, Matthew J. *Fatal Misconception: The Struggle to Control World Population.* Cambridge, Mass.: Harvard University Press, 2008.

Connolly, Cynthia A. *Saving Sickly Children: The Tuberculosis Preventorium in American Life, 1909–1970.* New Brunswick, N.J.: Rutgers University Press, 2008.

Coralita, Mary, Florence Boles, and Margaret Jacobsen. "Meeting a Polio Epidemic." *American Journal of Nursing* 53, 8 (August 1953): 935–938.

Cornely, Paul. "Polio Control—Ten Years on the March." *Opportunity* 26, 3 (1948): 111–120.

"County-Wide Use of Immune Globulin in the Modification and Prevention of Measles." *JAMA* 106 (May 1936): 1781–1783.

Cravens, Hamilton, Alan I. Marcus, and David M. Katzman, eds. *Technical Knowledge in American Culture: Science, Technology, and Medicine since the Early 1800s.* Tuscaloosa: University of Alabama Press, 1996.

Crawford, Dorothy H. *How Microbes Shaped Our History: Deadly Companions.* Oxford: Oxford University Press, 2007.

Creager, Angela N. H. *The Life of a Virus: Tobacco Mosaic Virus as an Experimental Model, 1930–1965.* Chicago: University of Chicago Press, 2002.

———. "'What Blood Told Dr Cohn': World War II, Plasma Fractionation, and the Growth of Human Blood Research." *Studies in History and Philosophy of Biological and Biomedical Sciences* 30, 3 (1999): 377–405.

Cueto, Marcos. *Missionaries of Science: The Rockefeller Foundation and Latin America.* Bloomington: Indiana University Press, 1994.

Cunningham, Andrew, and Perry Williams, eds. *The Laboratory Revolution in Medicine.* New York: Cambridge University Press, 1992.

Cutlip, Scott M. *Fund Raising in the United States: Its Role in America's Philanthropy.* Piscataway, N.J.: Transaction Publishers, 1990.

Daemmrich, Arthur, and Joanna Radin, eds. *Perspectives on Risk and Regulation: The FDA at 100.* Philadelphia: Chemical Heritage Foundation, 2007.

Daniel, Thomas M., and Frederick C. Robbins, eds. *Polio.* Rochester, N.Y.: University of Rochester Press, 1997.

David, H. "Le serum de convalescent dans la prophylaxie de la polymyelite." *Bulletin de l'Office Internationale de l'Hygiene Publique* 20 (1928).

Davis, Fred. *Passage through Crisis: Polio Victims and Their Families.* Indianapolis, Ind.: Bobbs-Merrill, 1963.

Davis, Lennard J., ed. *The Disability Studies Reader.* New York: Routledge, 1997.

Davis, Robert J. *The Healthy Skeptic: Cutting Through the Hype about Your Health.* Berkeley: University of California Press, 2008.

Davis-Floyd, Robbie, and James Dumit, eds. *Cyborg Babies: From Techno-Sex to Techno Tots.* New York: Routledge, 1998.

Dawson, Liza. "The Salk Polio Vaccine Trial of 1954: Risks, Randomization and Public Involvement in Research." *Clinical Trials* 1 (2004): 122–130.

Decker, Lewis G. *Images of America: Johnstown.* Charleston, S.C.: Arcadia Publishing, 1999.

Diedrich, Lisa. *Treatments: Language, Politics, and the Culture of Illness.* Minneapolis: University of Minnesota Press, 2007.

Doel, Ronald E., and Thomas Söderqvist, eds. *The Historiography of Contemporary Science, Technology, and Medicine: Writing Recent Science.* New York: Routledge, 2006.

Donat, James G. "Empirical Medicine in the 18th Century: The Rev. John Wesley's Search for Remedies That Work." *Methodist History* 44, 4 (2006): 216–226.

———. "The Rev. John Wesley's Extractions from Dr. Tissot: A Methodist Imprimatur." *Science History Publications* 39 (2001): 285–298.

Doorman S. J., ed. *Images of Science: Scientific Practice and the Public.* Brookfield, Vt.: Gower, 1989.

Douglas, Susan J. *Listening In: Radio and the American Imagination.* New York: Times Books, 1999.

Duffin, Jacalyn. *Lovers and Livers: Disease Concepts in History.* Toronto: University of Toronto Press, 2005.

Duffy, John. *From Humors to Medical Science: A History of American Medicine.* Urbana: University of Illinois Press, 1993.

———. *The Sanitarians: A History of American Public Health.* Urbana: University of Illinois Press, 1990.

Dulles, Foster Rhea. *The American Red Cross.* New York: Harper and Brothers, 1950.

Dutton, Paul V. *Differential Diagnosis: A Comparative History of Health Care Problems and Solutions in the United States and France*. Ithaca, N.Y.: Cornell University Press, 2007.

Dyck, Erika. *Psychedelic Psychiatry: LSD from Clinic to Campus*. Baltimore: Johns Hopkins University Press, 2008.

Dyck, Erika, and Christopher Fletcher, eds. *Locating Health*. London: Pickering and Chatto, 2011.

Edwards, Charles C. "Hearing Regulations and Regulations Describing Scientific Content of Adequate and Well-Controlled Scientific Investigations." *Federal Register* 35 (May 5, 1970): 7250–7253.

Elliott, Carl. *White Coat, Black Hat: Adventures on the Dark Side of Medicine*. Boston: Beacon Press, 2011.

Ellis, E. F., and C. S. Henney. "Adverse Reactions Following Administration of Human Gamma Globulin." *Journal of Allergy* 43, 1 (January 1969): 45–54.

Ellison, George T. H., Jay S. Kaufman, Rosemary F. Head, Paul A. Martin, and Jonathan D. Kahn. "Flaws in the U.S. Food and Drug Administration's Rationale for Supporting the Development and Approval of BiDil as a Treatment for Heart Failure Only in Black Patients." *Journal of Law, Medicine & Ethics* 36, 3 (2008): 449–457.

Emanuel, E., C. Grady, R. Crouch, R. Lie, F. Miller, and D. Wendler, eds. *Oxford Textbook of Research Ethics*. New York: Oxford University Press, 2008.

Etheridge, Elizabeth W. *Sentinel for Health: A History of the Centers for Disease Control*. Berkeley: University of California Press, 1992.

———. "Yellow Fever, Polio, and the New Public Health." *Reviews in American History* 21 (1993): 297–302.

Eyler, John M. "De Kruif's Boast: Vaccine Trials and the Construction of a Virus." *Bulletin of the History of Medicine* 80, 3 (2006): 409–438.

Fairchild, Amy L. "The Polio Narratives: Dialogues with FDR." *Bulletin of the History of Medicine* 75 (2001): 488–534.

Fairchild, Amy L., Ronald Bayer, and James Colgrove. *Searching Eyes: Privacy, the State, and Disease Surveillance in America*. Berkeley: University of California Press / New York: Milbank Memorial Fund, 2007.

Fairchild, Amy L., and Gerald M. Oppenheimer. "Public Health Nihilism vs. Pragmatism: History, Politics, and the Control of Tuberculosis." *AJPH* 88, 7 (July 1998): 1105–1117.

Farley, John. *To Cast Out Disease: A History of the International Health Division of the Rockefeller Foundation, 1913–1951*. New York: Oxford University Press, 2004.

Fauci, Anthony S., Eugene Braunwald, Dennis L. Kasper, Stephen L. Hauser, Dan L. Longo, J. Larry Jameson, and Joseph Loscalzo, eds. *Harrison's Principles of Internal Medicine*. 17th ed. New York: McGraw-Hill Medical Publishing Division, 2008.

Fee, Elizabeth, and Theodore M. Brown, eds. *Making Medical History: The Life and Times of Henry E. Sigerist*. Baltimore: Johns Hopkins University Press, 1997.

Ferrell, Robert H. *The Dying President: Franklin D. Roosevelt, 1944–1945*. Columbia: University of Missouri Press, 1998.

Fields, Barbara. "Ideology of Race in American History." In *Region, Race, and Reconstruction*, edited by J. Morgan Kousser and James M. McPherson. New York: Oxford University Press, 1982.

Finger, Anne. *Elegy for a Disease: A Personal and Cultural History of Polio*. New York: St. Martin's Press, 2006.

Fleck, Ludwik. *Genesis and Development of a Scientific Fact*. Translated by Fred Bradley and Thaddeus J. Trenn. Chicago: University of Chicago Press, 1979.

Flexner, Abraham. *Funds and Foundations: Their Policies Past and Present.* New York: Harper & Brothers, 1952.

Flexner, S., and H. D. Amos. "Localization of the Virus and Pathogenesis of Endemic Poliomyelitis." *Journal of Experimental Medicine* 20 (1914): 149–268.

Fosdick, Raymond B. *The Story of the Rockefeller Foundation.* New York: Harper & Brothers, 1952.

Foucault, Michel. *The Birth of the Clinic: An Archaeology of Medical Perception.* New York: Vintage Books, 1994.

———. *Discipline and Punish.* New York: Vintage Books, 1995.

Francis, Thomas, and Robert F. Korns. "An Evaluation of the 1954 Poliomyelitis Vaccine Trials: Summary Report." Ann Arbor, Mich.: Edwards Brothers, Inc., 1955.

Freudenburg, William R. "Seeding Science, Courting Conclusions: Reexamining the Intersection of Science, Corporate Cash, and the Law." *Sociological Forum* 20, 1 (March 2005): 3–33.

Freund, P. A., ed. *Experimentation with Human Subjects.* New York: George Braziller, 1970.

Freyhofer, Horst H. *The Nuremberg Medical Trial: The Holocaust and the Origin of the Nuremberg Medical Code.* New York: P. Lang, 2004.

Fuguitt, Glenn V., David L. Brown, and Calvin L. Beale. *Rural and Small Town America.* New York: Russell Sage Foundation, 1989.

Galambos, Louis, with Jane Eliot Sewell. *Networks of Innovation: Vaccine Development at Merck, Sharpe & Dohme, and Mulford, 1895–1995.* Cambridge: Cambridge University Press, 1997.

Galishoff, Stuart. "Germs Know No Color Line: Black Health and Public Policy in Atlanta, 1900–1918." *Journal of the History of Medicine* 40, 1 (1985): 22–41.

Gallagher, Hugh Gregory. *FDR's Splendid Deception.* New York: Dodd, Mead, 1985.

Gallagher, Richard B., Jean Gilder, G.J.V. Nossal, and Gaetano Salvatore, eds. *Immunology: The Making of a Modern Science.* London: Academic Press, Harcourt, Brace & Co., 1995.

Gallo, Robert. *Virus Hunting: AIDS, Cancer, and the Human Retrovirus: A Story of Scientific Discovery.* New York: Basic Books, 1991.

Gamble, Vanessa Northington. "Black Autonomy versus White Control: Black Hospitals and the Dilemmas of White Philanthropy, 1920–1940." *Minerva* 35 (1997): 247–267.

———. *Making a Place for Ourselves: The Black Hospital Movement.* New York: Oxford University Press, 1995.

———. "Under the Shadow of Tuskegee: African Americans and Health Care." *AJPH* 87, 11 (November 1997): 1773–1778.

"Gamma Globulin Effect on Poliomyelitis in 1953 Field Trial." *Public Health Reports* 69, 5 (May 1954): 519–520.

"Gamma Globulin in the Prophylaxis of Poliomyelitis." Public Health Monograph No. 20 / Public Health Service Publication No. 358. Washington, D.C.: U.S. Government Printing Office, 1954.

Gardner, Kirsten E. *Early Detection: Women, Cancer, and Awareness Campaigns in the Twentieth-Century United States.* Chapel Hill: University of North Carolina Press, 2006.

Gibson, Mary Eckenrode. "From Charity to an Able Body: The Care and Treatment of Disabled Children in Virginia, 1910–1935." PhD diss., University of Pennsylvania, 2007.

Gibson, Rosemary, and Janardan Prasad Singh. *Wall of Silence: The Untold Story of the Medical Mistakes That Kill and Injure Millions of Americans.* Washington, D.C.: Life-Line Press, 2003.

Gibson, Sam T. "Gamma Globulin." *American Journal of Nursing* 53, 6 (June 1953): 700–703.

Gilbert, James. *A Cycle of Outrage: America's Reaction to the Juvenile Delinquent in the 1950s.* New York: Oxford University Press, 1988.

Gilbert, Nigel G., and Michael Mulkay. *Opening Pandora's Box: A Sociological Analysis of Scientists' Discourse.* Cambridge: Cambridge University Press, 1984.

Gilbo, Patrick F. *The American Red Cross: The First Century.* New York: Harper & Row, 1981.

Gilman, Sander L. *Picturing Health and Illness: Images of Identity and Difference.* Baltimore: Johns Hopkins University Press, 1995.

Golden, Janet, Richard A. Meckel, and Heather Munro Prescott, eds. *Children and Youth in Sickness and in Health: A Historical Handbook and Guide.* Children and Youth: History and Culture Series. Westport, Conn.: Greenwood, 2004.

Golub, Edward S. *The Limits of Medicine: How Science Shapes Our Hope for the Cure.* Chicago: University of Chicago Press, 1997.

Gondola, Ch. Didier. *The History of Congo.* Westport, Conn.: Greenwood Press, 2002.

Goodman, Jordan, Anthony McElligott, and Lara Marks, eds. *Useful Bodies: Humans in the Service of Medical Science in the Twentieth Century.* Baltimore: Johns Hopkins University Press, 2003.

Gould, Tony. *A Summer Plague: Polio and Its Survivors.* New Haven: Yale University Press, 1995.

Graff, Harvey J. *Conflicting Paths: Growing up in America.* Cambridge, Mass.: Harvard University Press, 1995.

Gregory, Jane, and Steve Miller. *Science in Public: Communication, Culture, and Credibility.* New York: Plenum Trade, 1998.

Grey, Michael R. *New Deal Medicine: The Rural Health Programs of the Farm Security Administration.* Baltimore: Johns Hopkins University Press, 1999.

Grimshaw, M. L. "Scientific Specialization and the Poliovirus Controversy in the Years before World War II." *Bulletin of the History of Medicine* 69 (1995): 44–65.

Grob, Gerald N. *The Deadly Truth: A History of Disease in America.* Cambridge, Mass.: Harvard University Press, 2002.

Gross, Anne K. *The Polio Journals: Lessons from My Mother.* Greenwood Village, Colo.: Diversity Matters Press, 2011.

Gualde, Norbert. *Resistance: The Human Struggle against Infection.* Translated by Steven Rendall. Washington, D.C.: Dana Press, 2006.

Guerrini, Anita. *Experimenting with Humans and Animals: From Galen to Animal Rights.* Johns Hopkins Introductory Studies in the History of Science. Baltimore: Johns Hopkins University Press, 2003.

Haas, Jennifer Carrie. "Press Coverage of Three Epidemic Diseases: Influenza, Polio, and Measles." MA thesis, University of Minnesota, 2004.

Hall, Robert F. *Through the Storm: A Polio Story.* St. Cloud, Minn.: North Star Press, 1990.

Halliwell, Martin. "Cold War Ground Zero: Medicine, Psyops and the Bomb." *Journal of American Studies* 44, 2 (2007): 313–331.

———. *Therapeutic Revolutions: Medicine, Psychiatry, and American Culture, 1945–1970.* New Brunswick, N.J.: Rutgers University Press, 2014.

Halpern, Sydney Ann. *Lesser Harms: The Morality of Risk in Medical Research.* Chicago: University of Chicago Press, 2006.

Hammer, Patricia M. *The Decade of Elusive Promise: Professional Women in the United States, 1920–1930.* Ann Arbor, Mich.: University Microfilms International, 1979.

Hammon, W. M., L. L. Coriell, and J. Stokes Jr. "Evaluation of Red Cross Gamma Globulin as a Prophylactic Agent for Poliomyelitis: 1. Plan of Controlled Field Tests and Results of the 1951 Pilot Study in Utah." *JAMA* 150 (1952): 739–749.

———. "Evaluation of Red Cross Gamma Globulin as a Prophylactic Agent for Poliomyelitis: 2. Conduct and Early Followup of 1952 Texas and Iowa-Nebraska Studies." *JAMA* 150 (1952): 750–756.

Hammon, W. M., L. L. Coriell, P. F. Wehrle, C. R. Klimt, J. Stokes Jr. "Evaluation of Red Cross Gamma Globulin as a Prophylactic Agent for Poliomyelitis: 3. Preliminary Report of Results Based on Clinical Diagnoses." *JAMA* 150 (1952): 757–760.

Hammon, W. M., L. L. Coriell, P. F. Wehrle, J. Stokes Jr. "Evaluation of Red Cross Gamma Globulin as a Prophylactic Agent for Poliomyelitis: 4. Final Report of Results Based on Clinical Diagnoses." *JAMA* 151 (1953): 1272–1285.

Hammon, W. M., L. L. Coriell, E. H. Ludwig, R. M. McAllister, A. E. Greene, G. E. Sather, and P. F. Wehrle. "Evaluation of Red Cross Gamma Globulin as a Prophylactic Agent for Poliomyelitis. 5. Reanalysis of Results Based on Laboratory-Confirmed Cases," *JAMA* 156, 1 (1954): 21–27.

Hand, Douglas. "The Making of the Polio Vaccine." *American Heritage Magazine* 1, 1 (Summer 1985).

Hansen, Bert. *Picturing Medical Progress from Pasteur to Polio: A History of Mass Media Images and Popular Attitudes in America.* New Brunswick, N.J.: Rutgers University Press, 2009.

Harmon, F. Martin. *The Warm Springs Story: Legacy and Legend.* Macon, Ga.: Mercer University Press, 2014.

Harrington, Anne, ed. *The Placebo Effect: An Interdisciplinary Exploration.* Cambridge, Mass.: Harvard University Press, 1997.

Harrison, Helen E. "In the Picture of Health: Portraits of Health, Disease, and Citizenship in Canada's Public Health Advice Literature, 1920–1960." PhD diss., Queen's University, 2001.

Hartz, Jim, and Rick Chappell. *Worlds Apart: How the Distance between Science and Journalism Threatens America's Future.* Nashville, Tenn.: First Amendment Center, 1997.

Helfand, William H., Jan Lazarus, and Paul Theerman. "'. . . So That Others May Walk': The March of Dimes." *AJPH* 91, 8 (August 2001): 1190.

Hellman, Hal. *Great Feuds in Medicine: Ten of the Liveliest Disputes Ever.* New York: John Wiley & Sons, 2001.

Henle, Mary, ed. *Documents of Gestalt Psychology.* Berkeley: University of California Press, 1961.

Herman, Edward S., and Noam Chomsky. *Manufacturing Consent: The Political Economy of the Mass Media.* New York: Pantheon Books, 1988.

Hicks, Granville. *Small Town.* New York: Fordham University Press, 2004.

Hilts, Philip J. *Protecting America's Health: The FDA, Business, and One Hundred Years of Regulation.* Chapel Hill: University of North Carolina Press, 2003.

Hine, Darlene Clark. "Black Professionals and Race Consciousness: Origins of the Civil Rights Movement, 1890–1950." *Journal of American History* 89, 4 (March 2003): 1279–1294.

———. "The Corporeal and Ocular Veil: Dr. Matilda A. Evans (1872–1935) and the Complexity of Southern History." *Journal of Southern History* 70, 1 (2004): 3–34.

Hine, Thomas. *The Rise and Fall of the American Teenager*. New York: Bard, 1999.

Hoagwood, K., P. Jensen, and C. Fisher. *Ethical Issues in Mental Health Research with Children and Adolescents*. Mahwah, N.J.: Lawrence Erlbaum, 1996.

Hobsbawm, Eric, and Terence Ranger, eds. *The Invention of Tradition*. Cambridge: Cambridge University Press, 2010.

Holton, Gerald, and William A. Blanpied, eds. *Science and Its Public: The Changing Relationship*. Dordrecht, Holland: D. Reidel Pub. Co., 1976.

Hooper, Edward. *The River: A Journey to the Source of HIV and AIDS*. Boston: Little, Brown, 1999.

Hornblum, Allen M. *Acres of Skin: Human Experiments at Holmesburg Prison*. New York: Routledge, 1998.

———. *Sentenced to Science: One Black Man's Story of Imprisonment in America*. College Park: Pennsylvania State University Press, 2007.

Horstmann, Dorothy M. "The Poliomyelitis Story: A Scientific Hegira." *Yale Journal of Biology and Medicine* 58 (1985): 79–90.

———. "The Sabin Live Poliovirus Vaccine Trials in the USSR, 1959." *Yale Journal of Biology and Medicine* 64 (1991): 499–512.

Hughes, Thomas P. *American Genesis: A Century of Invention and Technological Enthusiasm, 1870–1970*. Chicago: University of Chicago Press, 2004.

Huisman, Frank, and John Harley Warner, eds. *Locating Medical History: The Stories and Their Meanings*. Baltimore: Johns Hopkins University Press, 2004.

Huse, Robert C. *Getting There: Growing Up with Polio in the 30's*. Bloomington, Ind.: 1stBooks, 2002.

Hutchinson, John F. *Champions of Charity: War and the Rise of the Red Cross*. Boulder, Colo.: Westview Press, 1996.

Illich, Ivan. *Medical Nemesis: The Expropriation of Health*. London: Calder & Boyars Ltd., 1975.

Imber, Jonathan B. *Trusting Doctors: The Decline of Moral Authority in American Medicine*. Princeton: Princeton University Press, 2008.

Ingraham, Hollis S. "Statistics and Medicine Knowledge." *AJPH* 48, 11 (November 1958): 1449–1459.

Ingstad, Benedicte, and Susan Reynolds Whyte, eds. *Disability and Culture*. Berkeley: University of California Press, 1995.

Irwin, Julia. *Making the World Safe: The American Red Cross and a Nation's Humanitarian Awakening*. New York: Oxford University Press, 2013.

Jackson, Mark. *Allergy: The History of a Modern Malady*. London: Reaktion Books, 2006.

Janeway, Charles A. *The Gamma Globulins*. Boston: Little, Brown, 1966.

Jannetta, Ann. *The Vaccinators: Smallpox, Medical Knowledge, and the "Opening" of Japan*. Stanford, Calif.: Stanford University Press, 2007.

Jenkins, Keith. *Re-thinking History*. New York: Routledge, 2003.

Jensen, Gwenn M. "System Failure: Health-Care Deficiencies in the World War II Japanese American Detention Centers." *Bulletin of the History of Medicine* 73, 4 (1999): 602–628.

Johnstone, D. *An Introduction to Disability Studies*. London: David Fulton Publishers, 1998.

Jones, Colin, and Roy Porter, eds. *Reassessing Foucault: Power, Medicine, and the Body*. Reprint. Studies in the Social History of Medicine. London: Routledge, 1998.

Jones, David S. *Rationalizing Epidemics: Meanings and Uses of American Indian Mortality since 1600.* Cambridge, Mass.: Harvard University Press, 2004.

Jones, James H. *Bad Blood: The Tuskegee Syphilis Experiment.* New York: The Free Press, 1981.

Jordanova, Ludmilla. *History in Practice.* London: Arnold Publishers, 2000.

Jurdem, Laurence R. "Return to the Arena: The Reemergence of Franklin D. Roosevelt, 1921–1928." MA thesis, University of Louisville, 1997.

Kallen, Stuart A. *The 1950s.* San Diego, Calif.: Greenhaven Press, 2000.

Kanigel, Robert. *The One Best Way: Frederick Winslow Taylor and the Enigma of Efficiency.* New York: Penguin-Viking, 1997.

Kasongo, Michael. *History of the Methodist Church in the Central Congo.* Lanham, Md.: University Press of America, 1998.

Keating, Peter, and Alberto Cambrosio. *Cancer on Trial: Oncology as a New Style of Practice.* Chicago: Chicago University Press, 2012.

Kehret, Peg. *Small Steps: The Year I Got Polio.* Morton Grove, Ill.: Albert Whitman, 1996.

Kellog, John. "Negro Urban Clusters in the Postbellum South." *Geographical Review* 67, 2 (1977): 166–180.

Kennett, Lee B. *G.I.: The American Soldier in World War II.* New York: Scribner, 1987.

Kett, J. F. *Rites of Passage: Adolescence in America.* New York: Basic Books, 1977.

Killalea, Anne. *The Great Scourge: The Tasmanian Infantile Paralysis Epidemic, 1937–1938.* Hobart, Tasmania: Tasmanian Historical Research Association, 1995.

Klein, Aaron E. *Trial by Fury: The Polio Vaccine Controversy.* New York: Charles Scribner's Sons, 1972.

Kloepping, Kent. *The Upside of the Downside: Journeys with a Companion Called Polio.* Tucson, Ariz.: Wheatmark Publishing, 2006.

Kluger, Jeffrey. *Splendid Solution: Jonas Salk and the Conquest of Polio.* New York: G. P. Putnam's Sons, 2004.

Kolchin, Peter. "Whiteness Studies: The New History of Race in America." *Journal of American History* 89, 1 (2002): 154–173.

Koprowski, Hilary, and Michael B. A. Oldstone, eds. *Microbe Hunters—Then and Now.* Bloomington, Ill.: Medi-Ed Press, 1996.

Krenn, Michael L., ed. *Race and U.S. Foreign Policy during the Cold War.* New York: Garland Publishing, 1998.

Kriegel, Leonard. *The Long Walk Home.* New York: Appleton-Century, 1964.

Kroker, Kenton, Jennifer Keelan, and Pauline M. H. Mazumdar, eds. *Crafting Immunity: Working Histories of Clinical Immunology.* Aldershot, UK: Ashgate Publishing, Ltd., 2008.

Krueger, Gretchen. *Hope and Suffering: Children, Cancer, and the Paradox of Experimental Medicine.* Baltimore: Johns Hopkins University Press, 2008.

Lafleur, William R., Gernot Böhme, and Susumu Shimazono, eds. *Dark Medicine: Rationalizing Unethical Medical Research.* Bioethics and the Humanities. Bloomington: Indiana University Press, 2007.

LaFollette, Marcel Chotkowski. *Making Science Our Own: Public Images of Science, 1910–1955.* Chicago: University of Chicago Press, 1990.

———. *Reframing Scopes: Journalists, Scientists, and Lost Photographs from the Trial of the Century.* Lawrence: University Press of Kansas, 2008.

Lambert, Sarah M. "Making History: Thomas Francis, Jr., MD, and the 1954 Salk Poliomyelitis Field Trial." *Archives of Pediatrics & Adolescent Medicine* 154 (May 2000): 512–517.

Larson, Richard C. "Perspectives on Queues: Social Justice and the Psychology of Queueing." *Operations Research* 35, 6 (November–December 1987): 895–905.

Latour, Bruno, and Steve Woolgar. *Laboratory Life: The Construction of Scientific Facts.* Princeton: Princeton University Press, 1986.

LaVeist, Thomas A, ed. *Race, Ethnicity, and Health.* San Francisco: Jossey-Bass, 2002.

Lax, Eric. *The Mould in Dr. Florey's Coat: The Story of the Penicillin Miracle.* New York: Henry Holt and Company, 2004.

Leavitt, Judith Walzer. *The Healthiest City: Milwaukee and the Politics of Health Reform.* Princeton: Princeton University Press, 1982.

———. *Make Room for Daddy: The Journey from Waiting Room to Birthing Room.* Chapel Hill: University of North Carolina Press, 2009.

———. *Typhoid Mary: Captive to the Public's Health.* Boston: Beacon Press, 1996.

Lederer, Susan E. *Flesh and Blood.* Oxford: Oxford University Press, 2008.

———. "'Porto Ricochet': Joking about Germs, Cancer, and Race Extermination in the 1930s." *American Literary History* 14, 4 (Winter 2002): 720–746.

———. *Subjected to Science: Human Experimentation in America before the Second World War.* Baltimore: Johns Hopkins University Press, 1997.

Lee, Sandra Soo-Jin, Joanna Mountain, and Barbara Koenig. "The Meanings of 'Race' in the New Genomics: Implications for Health Disparities Research." *Yale Journal of Health Policy, Law and Ethics* 1, 1 (2001): 33–75.

Leopold, Ellen. *Under the Radar: Cancer and the Cold War.* New Brunswick, N.J.: Rutgers University Press, 2009.

Lerner, Baron H. *Breast Cancer Wars: Hope, Fear, and the Pursuit of a Cure in Twentieth-Century America.* New York: Oxford University Press, 2001.

Lilienfeld, Abraham M. "Ceteris Paribus: The Evolution of the Clinical Trial." *Bulletin of the History of Medicine* 56 (Spring 1982): 1–18.

Lindee, M. Susan. *Suffering Made Real: American Science and the Survivors at Hiroshima.* Chicago: University of Chicago Press, 1994.

Linton, Simi. *Claiming Disability: Knowledge and Identity.* New York: New York University Press, 1998.

Lippman, Theo. *The Squire of Warm Springs: FDR in Georgia, 1924–1945.* Chicago: Playboy Press, 1978.

Lippmann, Walter. *Public Opinion.* New York: Macmillan, 1922.

Longmore, P. K., and L. Umansky, eds. *The New Disability History: American Perspectives.* New York: New York University Press, 2001.

Love, Spencie. *One Blood.* Chapel Hill: University of North Carolina Press, 1996.

Luce, R. Duncan, and Howard Raiffa. *Games and Decisions: Introduction and Critical Survey.* New York: Wiley, 1957.

MacDonald, Roger A. *A Country Doctor's Chronicle: Further Tales from the North Woods.* St. Paul: Minnesota Historical Society Press, 2004.

Magat, Richard. *The Ford Foundation at Work: Philanthropic Choices, Methods, and Styles.* New York: Plenum Press, 1979.

Magnello, Eileen, and Anne Hardy, eds. *The Road to Medical Statistics.* Clio Medica, vol. 67. Wellcome Series in the History of Medicine. Amsterdam: Editions Rodopi, 2002.

Mann, Leon. "Queue Culture: The Waiting Line as a Social System." *American Journal of Sociology* 75, 3 (November 1969): 340–354.

Marchand, Roland. *Advertising the American Dream: Making Way for Modernity, 1920–1940.* Berkeley: University of California Press, 1985.

Marks, Harry. "The 1954 Salk Poliomyelitis Vaccine Field Trial." *Clinical Trials* 8 (2011): 224–234.

———. *The Progress of Experiment: Science and Therapeutic Reform in the United States*, 1900–1990. Cambridge: Cambridge University Press, 1997, 2000.

Marsa, Linda. *Prescription for Profits: How the Pharmaceutical Industry Bankrolled the Unholy Marriage between Science and Business.* New York: Scribner, 1997.

Mawdsley, Stephen E. "Balancing Risks: Childhood Inoculations and America's Response to the Provocation of Paralytic Polio." *Social History of Medicine* 26, 4 (2013): 759–778.

———. "'Dancing on Eggs': Charles H. Bynum, Racial Politics, and the National Foundation for Infantile Paralysis, 1938–1954." *Bulletin of the History of Medicine* 84, 2 (Summer 2010): 217–247.

———. "Harnessing the Power of People: The Fundraising Efforts of the National Foundation for Infantile Paralysis, 1938–1945." BA honors thesis, University of Alberta, 2006.

———. "Polio and Prejudice: Charles Hudson Bynum and the Racial Politics of the National Foundation for Infantile Paralysis, 1938–1954." MA thesis, University of Alberta, 2008.

Maxwell, J. H. "The Iron Lung: Halfway Technology or Necessary Step?" *Milbank Quarterly* 64 (1986): 3–33.

May, Elaine Tyler. *Homeward Bound: American Families in the Cold War.* New York: Basic Books, 1990.

Mazumdar, Pauline M. H., ed. *Immunology, 1930–1980: Essays on the History of Immunology.* Toronto: Wall and Thompson, 1989.

McBean, Eleanor. *The Poisoned Needle: Suppressed Facts about Vaccination.* Pomeroy, Wash.: Health Research, 1993.

McKhann, Charles F. "The Prevention and Modification of Measles." *JAMA* 109 (December 1937): 2034–2038.

McPhee, Stephen J., and Maxine A. Papadakis, eds. *CURRENT Medical Diagnosis and Treatment 2009.* New York: McGraw-Hill Medical Publishing Division, 2008.

Meader, F. M. "Scarlet Fever Prophylaxis: Use of Blood Serum from Persons Who Have Recovered from Scarlet Fever." *JAMA* 94 (March 1930): 622–625.

Meier, Paul. "The Biggest Public Health Experiment Ever: The 1954 Field Trial of the Salk Poliomyelitis Vaccine." In *Statistics: A Guide to the Unknown*, edited by F. Mosteller et al., 2–13. San Francisco: Holden-Day, 1972.

———. "Statistics and Medical Experimentation." *Biometrics* 31 (June 1975): 511–529.

Meldrum, Marcia Lynn. "'A Calculated Risk': The Salk Polio Vaccine Field Trials of 1954." *British Medical Journal* 317 (October 31, 1998): 1233–1236.

———. "Departures from the Design: The Randomized Clinical Trial in Historical Context, 1946–1970." PhD diss., State University of New York, 1994.

Mendelsohn, Robert S. *Confessions of a Medical Heretic.* Chicago: Contemporary Books, 1979.

Merton, Robert K. *Mass Persuasion: The Social Psychology of a War Bond Drive.* New York: Harper & Brothers, 1946.

———. *The Sociology of Science: Theoretical and Empirical Investigations.* Chicago: University of Chicago Press, 1979.

Milgram, Stanley. *Obedience to Authority: An Experimental View.* New York: Harper & Row, 1974.

Mintz, Steven. *Huck's Raft: A History of American Childhood.* Cambridge, Mass.: Belknap Press of Harvard University Press, 2004.

Miracle at Hickory. New York: National Foundation for Infantile Paralysis, 1944.

Moldow, Gloria. *Women Doctors in Gilded-Age Washington*. Urbana: University of Illinois Press, 1987.

Monto, Arnold S. "Francis Field Trial of Inactivated Poliomyelitis Vaccine: Background and Lessons for Today." *Epidemiological Reviews* 21, 1 (1999): 7–23.

Morantz-Sanchez, Regina. *Sympathy and Science: Women Physicians in American Medicine*. New York: Oxford University Press, 1985.

Moreno, Jonathan D. *Undue Risk: Secret State Experiments on Humans*. New York: Routledge, 2001.

Moreno, Jonathan D., ed. *In the Wake of Terror: Medicine and Morality in a Time of Crisis*. Cambridge, Mass.: MIT Press, 2003.

Morris, Joan E. *Polio & Me, Now & Then: Now Is 2004—Then Was 1942*. Bloomington, Ind.: AuthorHouse, 2004.

Mortimer, Robert P. "Convalescent Whole Blood, Plasma, and Serum in the Prophylaxis of Measles." *Reviews in Medical Virology* 15 (2005): 407–421.

Mosteller, Frederick. "Gamma Globulin in the Prophylaxis of Poliomyelitis: An Evaluation of the Efficacy of Gamma Globulin in the Prophylaxis of Paralytic Poliomyelitis as Used in the United States 1953." *Journal of the American Statistical Association* 49, 268 (December 1954): 926–927.

Moynihan, Ray, and Alan Cassels. *Selling Sickness: How the World's Biggest Pharmaceutical Companies Are Turning Us All into Patients*. New York: Perseus Books Group, 2005.

Mullally, Sasha. *Unpacking the Black Bag: A History of North American Country Doctors, 1900–1950*. Toronto: University of Toronto Press, 2009.

Murcott, Ann, ed. *Sociology and Medicine*. Burlington, Vt.: Ashgate Publishing, 2006.

Needell, Allan A. *Science, Cold War, and the American State*. New York: Routledge, 2012.

Nielsen, Waldemar A. *The Golden Donors: A New Anatomy of the Great Foundations*. New York: Truman Talley Books / E. P. Dutton, 1985, 1989.

Nordenberg, Mark A. *Defeat of an Enemy: Chancellor Mark A. Nordenberg Reports on the 50th Anniversary Celebration of the Triumph of the Pitt Polio Vaccine*. Pittsburgh: University of Pittsburgh, 2005.

Oakes, Guy. *The Imaginary War: Civil Defense and Cold War Culture*. Oxford: Oxford University Press, 1994.

O'Donnell, John. *Coriell: The Coriell Institute for Medical Research and a Half Century of Science*. Sagamore Beach, Mass.: Watson Publishing International, Science History Publications Series, 2002.

Offit, Paul A. *The Cutter Incident: How America's First Polio Vaccine Led to the Growing Vaccine Crisis*. New Haven: Yale University Press, 2005.

———. *Vaccinated: One Man's Quest to Defeat the World's Deadliest Diseases*. Washington, D.C.: Smithsonian, 2007.

Oreskes, Naomi, and Erik M. Conway. *Merchants of Doubt: How a Handful of Scientists Obscured the Truth on Issues from Tobacco Smoke to Global Warming*. New York: Bloomsbury, 2010.

Oshinsky, David M. *A Conspiracy So Immense: The World of Joe McCarthy*. New York: Oxford University Press, 2005.

———. *Polio: An American Story: The Crusade That Mobilized the Nation against the 20th Century's Most Feared Disease*. Oxford: Oxford University Press, 2005.

Ostling, Richard N., and Joan K. Ostling. *Mormon America: The Power and the Promise.* New York: HarperOne, 2007.

Pakenham, Thomas. *The Scramble for Africa: White Man's Conquest of the Dark Continent, 1879 to 1912.* New York: Avon Books, 1992.

Page, Benjamin B., and David A. Valone, eds. *Philanthropic Foundations and the Globalization of Scientific Medicine and Public Health.* Lanham, Md.: University Press of America, 2007.

Paul, John R. *A History of Poliomyelitis.* New Haven: Yale University Press, 1971.

Perdue, Thomas R. "Poliomyelitis Vaccine: A Report on Field Trials in Southern Alameda County in 1954." *California Medicine* 83, 3 (September 1955): 233–236.

Perks, Robert, and Alistair Thomson. *The Oral History Reader.* London: Routledge, 1998.

Perlich, Pamela S. *Utah Minorities: The Story Told by 150 Years of Census Data.* Bureau of Economic and Business Research, David S. Eccles School of Business, University of Utah, October 2002.

Pernick, Martin S. *The Black Stork: Eugenics and the Death of "Defective" Babies in American Medicine and Motion Pictures since 1915.* New York: Oxford University Press, 1996.

———. *A Calculus of Suffering: Pain, Professionalism, and Anesthesia in Nineteenth-Century America.* New York: Columbia University Press, 1985.

Peters, Toine. *Interferon: The Science and Selling of a Miracle Drug.* Routledge Studies in the History of Science, Technology, and Medicine, no. 21. London: Routledge, 2005.

Phillips, Joseph Michael. "The File This Time: The Battle over Racial, Regional, and Religious Identities in Dallas, Texas, 1860–1990." PhD diss., University of Texas at Austin, 2002.

Pinch, Trevor. "'Testing—One, Two, Three . . . Testing!': Toward a Sociology of Testing." *Science, Technology, & Human Values* 18, 1 (Winter 1993): 25–41.

Plagemann, Bentz. *My Place to Stand.* New York: Farrar, Straus, 1949.

Porter, Roy. *Blood and Guts: A Short History of Medicine.* New York: W. W. Norton, 2003.

———. *The Cambridge Illustrated History of Medicine.* Cambridge: Cambridge University Press, 1996.

———. "The Patients' View: Doing Medical History from Below." *Theory and Society* 14, 2 (March 1985): 175–198.

Porter, Theodore M. *Trust in Numbers: The Pursuit of Objectivity in Science and Public Life.* Princeton: Princeton University Press, 1995.

Powell, Allen Kent, ed. *Utah History Encyclopedia.* Salt Lake City: University of Utah Press, 1994.

"Practical Experience with Poliomyelitis Vaccine: Questions and Answers." *AJPH* 46 (May 1956): 563–574.

Reagan, Leslie J., Nancy Tomes, and Paula A. Treichler, eds. *Medicine's Moving Pictures: Medicine, Health, and Bodies in American Film and Television.* Rochester Studies in Medical History. Rochester, N.Y.: University of Rochester Press, 2007.

Reese, W. *The Origins of the American High School.* New Haven: Yale University Press, 1995.

Reverby, Susan M. *Examining Tuskegee.* Chapel Hill: University of North Carolina Press, 2009.

———. "'Normal Exposure' and Inoculation Syphilis: A PHS 'Tuskegee' Doctor in Guatemala, 1946–48." *Journal of Policy History* 23, 1 (January 2011): 6–28.

Reverby, Susan M., ed. *Tuskegee's Truths: Rethinking the Tuskegee Syphilis Study.* Chapel Hill: University of North Carolina Press, 2000.

Rinaldo, Charles R. "Passive Immunization against Poliomyelitis: The Hammon Gamma Globulin Field Trials, 1951–1953." *AJPH* 95, 5 (May 2005): 790–799.

Risse, G. B. "Revolt against Quarantine: Community Responses to the 1916 Polio Epidemic, Oyster Bay, New York." *Transactions & Studies of the College of Physicians of Philadelphia* 14 (1992): 23–50.

Ritchie, Donald D. *Doing Oral History: A Practical Guide.* New York: Oxford University Press, 2003.

Robbins, Frederick C. "The History of Polio Vaccine Development." In *Vaccines*, 4th ed., edited by Stanley A. Plotkin and Walter A. Orenstein. Philadelphia: Elsevier, 2004.

Robinson, Judith. *Noble Conspirator: Florence S. Mahoney and the Rise of the National Institutes of Health.* Washington, D.C.: Francis Press, 2001.

Roelcke, Volker, and Giovanni Maio, eds. *Twentieth-Century Ethics of Human Subjects Research: Historical Perspectives on Values, Practices, and Regulations.* Medical History. Stuttgart: Franz Steiner, 2004.

Rogers, Naomi. *Dirt and Disease: Polio before FDR.* New Brunswick, N.J.: Rutgers University Press, 1992.

———. *Polio Wars: Sister Kenny and the Golden Age of American Medicine.* New York: Oxford University Press, 2014.

———. "Race and the Politics of Polio: Warm Springs, Tuskegee, and the March of Dimes." *AJPH* 97, 4 (May 2007): 784–795.

———. "Screen the Baby, Swat the Fly: Polio in the Northeastern United States, 1916." PhD diss., University of Pennsylvania, 1986.

———. "What Is Scientific Evidence?: Sister Kenny, American Doctors, and Polio Therapy, 1940–1952." In *"Outpost Medicine": Australian Studies on the History of Medicine.* Hobart: Australian Society of the History of Medicine, 1994.

Rogers, Robert F. *Destiny's Landfall: A History of Guam.* Honolulu: University of Hawai'i Press, 1995.

Rose, David W. *Images of America: March of Dimes.* Charleston, S.C.: Arcadia Publishing, 2003.

Rosen, George. *A History of Public Health.* Baltimore: Johns Hopkins University Press, 1993.

Rosenberg, Charles E., and Janet Golden, eds. *Explaining Epidemics and Other Studies in the History of Medicine.* Cambridge: Cambridge University Press, 1992.

———. *Framing Disease: Studies in Cultural History.* New Brunswick, N.J.: Rutgers University Press, 1992.

———. *No Other Gods: On Science and American Social Thought.* Baltimore: Johns Hopkins University Press, 1997.

———. *Our Present Complaint: American Medicine, Then and Now.* Baltimore: Johns Hopkins University Press, 2007.

Ross, Walter S. *Crusade: The Official History of the American Cancer Society.* New York: Arbor House, 1987.

Rossiter, Margaret. *Women Scientists in America: Struggles and Strategies to 1940.* Baltimore: Johns Hopkins University Press, 1982.

Roth, Phillip. *Nemesis.* London: Jonathan Cape, 2010.

Rothstein, William G. *Public Health and the Risk Factor: A History of an Uneven Medical Revolution.* Rochester, N.Y.: University of Rochester Press, 2003.

Rowland, John. *The Polio Man: The Story of Dr. Jonas Salk.* New York: Roy Publishers, 1960.

Rushefsky, Mark E., and Deborah R. McFarlane. *The Politics of Public Health in the United States*. Armonk, N.Y.: M. E. Sharpe, 2005.

Rutty, Christopher J. "Do Something! . . . Do Anything!: Poliomyelitis in Canada, 1927–1962." PhD diss., University of Toronto, 1995.

———. "The Middle-Class Plague: Epidemic Polio and the Canadian State, 1936–1937." *Canadian Bulletin of Medical History* 13, 2 (1996): 277–314.

Said, Edward. *Orientalism*. New York: Vintage Books, 1994.

Sako, Wallace, P. F. Dwan, and E. S. Platou. "Sulfanilamide and Serum in the Treatment and Prophylaxis of Scarlet Fever." *JAMA* 111 (September 1938): 995–997.

Samuel, Lawrence R. *Pledging Allegiance: American Identity and the Bond Drive of World War II*. Washington, D.C.: Smithsonian Institution Press, 1991.

Sandler, Benjamin P. *Diet Prevents Polio*. Milwaukee, Wis.: Lee Foundation for Nutritional Research, 1951.

Sass, Edmund, George Gottfried, and Anthony Soren. *Polio's Legacy: An Oral History*. Lanham, Md.: University Press of America, 1996.

Savage, Jon. *Teenage: The Creation of Youth Culture*. New York: Viking, 2007.

Schoen, Johanna. *Choice and Coercion: Birth Control, Sterilization, and Abortion in Public Health and Welfare*. Chapel Hill: University of North Carolina Press, 2005.

Scott, James C. *Seeing Like a State: How Certain Schemes to Improve the Human Condition Have Failed*. New Haven: Yale University Press, 1998.

Scull, Andrew. *Madhouse: A Tragic Tale of Megalomania and Modern Medicine*. New Haven: Yale University Press, 2005.

Sealander, Judith. *Private Wealth and Public Life: Foundation Philanthropy and the Reshaping of American Social Policy from the Progressive Era to the New Deal*. Baltimore: Johns Hopkins University Press, 1997.

Seavey, Nina G., Jane S. Smith, Paul Wagner, eds. *A Paralyzing Fear: The Triumph over Polio in America*. New York: TV Books, 1998.

Serotte, Brenda. *The Fortune Teller's Kiss*. Lincoln: University of Nebraska Press, 2006.

Seytre, Bernard, and Mary Shaffer. *The Death of a Disease: A History of the Eradication of Poliomyelitis*. New Brunswick, N.J.: Rutgers University Press, 2005.

Shah, Sonia. *The Body Hunters: Testing New Drugs on the World's Poorest Patients*. New York: The New Press, 2007.

Shapiro, Arthur K., and Elaine Shapiro. *The Powerful Placebo: From Ancient Priest to Modern Physician*. Baltimore: Johns Hopkins University Press, 1997.

Sharma, Ram Nath. *Experimental Psychology*. New Delhi: Atlantic Publishers & Distributors, 2006.

Shell, Marc. *Polio and Its Aftermath: The Paralysis of Culture*. Cambridge, Mass.: Harvard University Press, 2005.

Shorter, Edward. *The Health Century*. New York: Doubleday, 1987.

Showalter, Elaine. *Hystories: Hysterical Epidemics and Modern Media*. New York: Columbia University Press, 1997.

Shreve, Susan Richards. *Warm Springs: Traces of a Childhood at FDR's Polio Haven*. Boston: Houghton Mifflin Company, 2007.

Sieber, Joan. *Planning Ethically Responsible Research: A Guide for Students and Internal Review Boards*. Newbury Park, Calif.: Sage Publications, 1992.

Sills, David L. *The Volunteers: Means and Ends in a National Organization*. Glencoe, Ill.: Free Press, 1957.

Silverstein, Arthur M. *Paul Ehrlich's Receptor Immunology: The Magnificent Obsession*. San Diego, Calif.: Academic Press, 2002.

Skloot, Rebecca. *The Immortal Life of Henrietta Lacks.* New York: Macmillan, 2010.
Smith, Jane S. *Patenting the Sun: Polio and the Salk Vaccine.* New York: William Morrow, 1990.
Smith, Susan L. *Japanese American Midwives: Culture, Community, and Health Politics, 1880–1950.* Urbana: University of Illinois Press, 2005.
———. *Sick and Tired of Being Sick and Tired: Black Women's Health Activism in America, 1890–1950.* Philadelphia: University of Pennsylvania Press, 1995.
Snyder, Sharon L., and David T. Mitchell. *Cultural Locations of Disability.* Chicago: University of Chicago Press, 2006.
Spitz, Vivien. *Doctors from Hell: The Horrific Account of Nazi Experiments on Humans.* Boulder, Colo.: Sentient Publications, 2005.
Stanton, Jennifer, ed. *Innovations in Health and Medicine: Diffusion and Resistance in the Twentieth Century.* London: Routledge, 2002.
Stapleton, Darwin H., ed. *Creating a Tradition of Biomedical Research: Contributions to the History of the Rockefeller University.* New York: Rockefeller University Press, 2004.
Starr, Douglas. *Blood: An Epic History of Medicine and Commerce.* New York: Alfred A. Knopf, 1998.
Starr, Paul. *The Social Transformation of American Medicine.* New York: Basic Books, 1982.
Stephens, Martha. *The Treatment: The Story of Those Who Died in the Cincinnati Radiation Tests.* Durham, N.C.: Duke University Press, 2002.
Sterling, Dorothy. *Polio Pioneers: The Story of the Fight against Polio.* Garden City, N.Y.: Doubleday, 1955.
Stern, Alexandra Minna, and Howard Markel, eds. *Formative Years: Children's Health in the United States, 1880–2000.* Ann Arbor: University of Michigan Press, 2002.
Stevens, M. L. Tina. *Bioethics in America: Origins and Cultural Politics.* Baltimore: Johns Hopkins University Press, 2000.
Stevens, Rosemary A. *American Medicine and the Public Interest: A History of Specialization.* Berkeley: University of California Press, 1998.
———. *The Public-Private Health Care State: Essays on the History of American Health Care Policy.* New Brunswick, N.J.: Transaction Publishers, 2007.
Stevens, Rosemary A., Charles E. Rosenberg, and Lawton R. Burns, eds. *History and Health Policy in the United States: Putting the Past Back In.* New Brunswick, N.J.: Rutgers University Press, 2006.
Strathern, Paul. *Medicine: From Hippocrates to Gene Therapy.* London: Robinson, 2005.
Strauss, Maurice B., ed. *Familiar Medical Quotations.* Boston: Little, Brown, 1968.
Sugrue, Thomas J. *The Origins of the Urban Crisis: Race and Inequality in Postwar Detroit.* Princeton: Princeton University Press, 2005.
Surgenor, Douglas M. *Edwin J. Cohn and the Development of Protein Chemistry: With a Detailed Account of His Work on the Fractionation of Blood during and after World War II.* Cambridge, Mass.: Harvard University Press, 2002.
Taub S. J. "Adverse Reactions Following Administration of Human Gamma Globulin." *Eye, Ear, Nose & Throat Monthly* 48, 6 (June 1969): 393–394.
Thompson, Paul. *The Voice of the Past: Oral History.* Oxford: Oxford University Press, 1988.
Tomes, Nancy. *The Gospel of Germs: Men, Women, and the Microbe in American Life.* Cambridge: Cambridge University Press, 1998.

Tone, Andrea, and Elizabeth Siegel Watkins, eds. *Medicating Modern America: Prescription Drugs in History.* New York: New York University Press, 2007.

Toumey, Christopher P. *Conjuring Science: Scientific Symbols and Cultural Meanings in American Life.* New Brunswick, N.J.: Rutgers University Press, 1996.

Troan, John. *Passport to Adventure: Or, How a Typewriter from Santa Led to an Exciting Lifetime Journey.* Pittsburgh: Cold-Comp, 2000.

Trotter, Griffin. *The Ethics of Coercion in Mass Casualty Medicine.* Baltimore: Johns Hopkins University Press, 2007.

Turner, David M., and Kevin Stagg, eds. *Social Histories of Disability and Deformity.* London: Routledge, 2006.

Tushnet, Leonard. *The Medicine Men: The Myth of Quality Medical Care in America Today.* New York: St. Martin's Press, 1971.

Ueda, Reed. *Avenues to Adulthood: The Origins of the High School and Social Mobility in an American Suburb.* Cambridge: Cambridge University Press, 1987.

Umbach, Günter. *Successfully Marketing Clinical Trial Results: Winning in the Healthcare Business.* Farnham, UK: Gower Publishing, 2006.

"Unsung Hero of the War on Polio." *Public Health Magazine.* University of Pittsburgh Graduate School of Public Health, 2004.

Vaughan, Roger. *Listen to the Music: The Life of Hilary Koprowski.* New York: Springer-Verlag, 2000.

Vidich, Arthur J., and Joseph Bensman. *Small Town in Mass Society: Class, Power, and Religion in a Rural Community.* Princeton: Princeton University Press, 1968.

Wailoo, Keith. *Drawing Blood: Technology and Disease Identity in Twentieth-Century America.* Baltimore: Johns Hopkins University Press, 1997.

———. *Dying in the City of the Blues: Sickle Cell Anemia and the Politics of Race and Health.* Chapel Hill: University of North Carolina Press, 2001.

Wailoo, Keith, and Stephen Pemberton. *The Troubled Dream of Genetic Medicine: Ethnicity and Innovation in Tay-Sachs, Cystic Fibrosis, and Sickle Cell Disease.* Baltimore: Johns Hopkins University Press, 2006.

Walker, Turnley. *Rise Up And Walk.* New York: E. P. Dutton, 1950.

Walters, Ronald G. *Scientific Authority and Twentieth-Century America.* Baltimore: Johns Hopkins University Press, 1997.

Wang, Caroline C. "Portraying Stigmatized Conditions: Disabling Images in Public Health." *Journal of Health Communication* 3, 2 (May 1998): 149–159.

Wang, Jessica. *American Science in an Age of Anxiety: Scientists, Anticommunism, and the Cold War.* Chapel Hill: University of North Carolina Press, 1999.

Wang, Thomas Hsing-Teh. "Gamma Globulin: The Development and Implementation of a Polio Preventive in the 1950s." AB honors thesis, Harvard University, 1997.

Ward, John W., and Christian Warren, eds. *Silent Victories: The History and Practice of Public Health in Twentieth-Century America.* Oxford: Oxford University Press, 2007.

Ward, Susan Mechele. "Rhetorically Constructing a 'Cure': FDR's Dynamic Spectacle of Normalcy." PhD diss., Regent University, 2005.

Warwick, William. "Convalescent Serum in the Prevention of Measles." *Canadian Medical Association Journal* 21, 6 (December 1929): 694–696.

Watts, Sheldon. *Epidemics and History: Disease, Power, and Imperialism.* New Haven: Yale University Press, 1997.

Weaver, George H., and T. T. Crooks. "The Use of Convalescent Serum in the Prophylaxis of Measles." *JAMA* 82 (January 1924): 204–206.

Welsome, Eileen. *The Plutonium Files: America's Secret Medical Experiments in the Cold War*. New York: Dial Press, 1999.

Wesley, John. *Primitive Physic; or, An Easy and Natural Method of Curing Most Diseases*. Bristol: William Pine, 1773.

Weyers, Wolfgang. *The Abuse of Man: An Illustrated History of Dubious Medical Experimentation*. New York: Ardor Scribendi, 2003.

Wheatley, Steven C. *The Politics of Philanthropy: Abraham Flexner and Medical Education*. Madison: University of Wisconsin Press, 1988.

Williams, Carol Grace. "Health Conceptions and Practices of Individuals Who Have Had Poliomyelitis." PhD diss., Boston University, 1990.

Williams, Gareth. *Paralysed with Fear: The Story of Polio*. New York: Palgrave Macmillan, 2013.

Wilson, Daniel J. "Basil O'Connor, the National Foundation for Infantile Paralysis and the Reorganization of Polio Research in the United States, 1935–41." *Journal of the History of Medicine and Allied Sciences* (March 2014) doi:10.1093/jhmas/jru003.

———. "A Crippling Fear: Experiencing Polio in the Era of FDR." *Bulletin of the History of Medicine* 72 (1998): 464–495.

———. *Living with Polio: The Epidemic and Its Survivors*. Chicago: University of Chicago Press, 2005.

———. *Polio*. Westport, Conn.: Greenwood Publishing Group, 2009.

———. "Psychological Trauma and Its Treatment in the Polio Epidemics." *Bulletin of the History of Medicine* 82, 4 (Winter 2008): 848–877.

Wilson, J. "Sister Elizabeth Kenny's Trial by the Royal Commission." *History of Nursing Society Journal* 4 (1992–1993): 91–99.

Wilson, John Rowan. *Margin of Safety: The Story of Poliomyelitis Vaccine*. London: Collins Clear-Type Press, 1963.

Winkelstein, Warren, and Saxon Graham. "Factors in Participation in the 1954 Poliomyelitis Vaccine Field Trials, Erie County, New York." *AJPH* 49, 11 (November 1959): 1454–1466.

Winkler, Allan M. *The Politics of Propaganda: The Office of War Information, 1942–1945*. New Haven: Yale University Press, 1978.

Wintrobe, Maxwell M. *Blood, Pure and Eloquent: A Story of Discovery, of People, and of Ideas*. New York: McGraw-Hill, 1980.

Woods, Regina. *Tales from Inside the Iron Lung: And How I Got Out of It*. Philadelphia: University of Pennsylvania Press, 1994.

Woodward, Theodore E. *The Armed Forces Epidemiological Board: Its First Fifty Years*. Falls Church, Va.: Office of the Surgeon General, 1990.

Wooten, Heather Green. "The Polio Years in Harris and Galveston Counties, 1930–1962." PhD diss., University of Texas at Galveston, 2006.

———. *The Polio Years in Texas: Battling a Terrifying Unknown*. College Station: Texas A&M University Press, 2009.

Wyatt, H. V. "Provocation of Poliomyelitis by Multiple Injections." *Transactions of the Royal Society of Tropical Medicine and Hygiene* 79, 3 (1985): 355–358.

———. "Provocation Poliomyelitis: Neglected Clinical Observations from 1914 to 1950." *Bulletin of the History of Medicine* 55 (1981): 543–537.

Young, Paul Thomas. *Motivation of Behavior: The Fundamental Determinants of Human and Animal Activity*. New York: J. Wiley & Sons, 1936.

Zhou, Rongrong, and Dilip Soman. "Looking Back: Exploring the Psychology of Queuing and the Effect of the Number of People Behind." *Journal of Consumer Research: An Interdisciplinary Quarterly* 29, 4 (March 2003): 517–530.

Zingher, Abraham. "Convalescent Whole Blood, Plasma, and Serum in Prophylaxis of Measles." *JAMA* 82, 15 (1924): 1180–1187.

———. "The Use of Convalescent and Normal Blood in the Treatment of Scarlet Fever." *JAMA* 65, 10 (1915): 875–877.

Index

About the Author

Stephen E. Mawdsley is a social historian of twentieth-century American medicine. His research has explored African American health activism, societal perceptions of vaccination, and the development of health charities. In 2012, after earning degrees at the University of Alberta and the University of Cambridge, Stephen became the Isaac Newton–Ann Johnston Research Fellow at Clare Hall, University of Cambridge. He took up a Wellcome Trust University Award at the University of Strathclyde in 2016.

Available titles in the Critical Issues in Health and Medicine series:

Emily K. Abel, *Suffering in the Land of Sunshine: A Los Angeles Illness Narrative*

Emily K. Abel, *Tuberculosis and the Politics of Exclusion: A History of Public Health and Migration to Los Angeles*

Marilyn Aguirre-Molina, Luisa N. Borrell, and William Vega, eds. *Health Issues in Latino Males: A Social and Structural Approach*

Anne-Emanuelle Birn and Theodore M. Brown, eds., *Comrades in Health: U.S. Health Internationalists, Abroad and at Home*

Susan M. Chambré, *Fighting for Our Lives: New York's AIDS Community and the Politics of Disease*

James Colgrove, Gerald Markowitz, and David Rosner, eds., *The Contested Boundaries of American Public Health*

Cynthia A. Connolly, *Saving Sickly Children: The Tuberculosis Preventorium in American Life, 1909–1970*

Tasha N. Dubriwny, *The Vulnerable Empowered Woman: Feminism, Postfeminism, and Women's Health*

Edward J. Eckenfels, *Doctors Serving People: Restoring Humanism to Medicine through Student Community Service*

Julie Fairman, *Making Room in the Clinic: Nurse Practitioners and the Evolution of Modern Health Care*

Jill A. Fisher, *Medical Research for Hire: The Political Economy of Pharmaceutical Clinical Trials*

Alyshia Gálvez, *Patient Citizens, Immigrant Mothers: Mexican Women, Public Prenatal Care and the Birth Weight Paradox*

Gerald N. Grob and Howard H. Goldman, *The Dilemma of Federal Mental Health Policy: Radical Reform or Incremental Change?*

Gerald N. Grob and Allan V. Horwitz, *Diagnosis, Therapy, and Evidence: Conundrums in Modern American Medicine*

Rachel Grob, *Testing Baby: The Transformation of Newborn Screening, Parenting, and Policymaking*

Mark A. Hall and Sara Rosenbaum, eds., *The Health Care "Safety Net" in a Post-Reform World*

Laura D. Hirshbein, *American Melancholy: Constructions of Depression in the Twentieth Century*

Laura Hirshbein, *Smoking Privileges: Psychiatry, the Mentally Ill, and the Tobacco Industry in America*

Timothy Hoff, *Practice under Pressure: Primary Care Physicians and Their Medicine in the Twenty-first Century*

Beatrix Hoffman, Nancy Tomes, Rachel N. Grob, and Mark Schlesinger, eds., *Patients as Policy Actors*

Ruth Horowitz, *Deciding the Public Interest: Medical Licensing and Discipline*

Rebecca M. Kluchin, *Fit to Be Tied: Sterilization and Reproductive Rights in America, 1950–1980*

Jennifer Lisa Koslow, *Cultivating Health: Los Angeles Women and Public Health Reform*

Susan C. Lawrence, Privacy and the Past: Research, Law, Archives, Ethics

Bonnie Lefkowitz, *Community Health Centers: A Movement and the People Who Made It Happen*

Ellen Leopold, *Under the Radar: Cancer and the Cold War*

Barbara L. Ley, *From Pink to Green: Disease Prevention and the Environmental Breast Cancer Movement*

Sonja Mackenzie, *Structural Intimacies: Sexual Stories in the Black AIDS Epidemic*

Stephen E. Mawdsley, Selling Science: Polio and the Promise of Gamma Globulin

David Mechanic, *The Truth about Health Care: Why Reform Is Not Working in America*

Richard A. Meckel, *Classrooms and Clinics: Urban Schools and the Protection and Promotion of Child Health, 1870–1930*

Alyssa Picard, *Making the American Mouth: Dentists and Public Health in the Twentieth Century*

Heather Munro Prescott, *The Morning After: A History of Emergency Contraception in the United States*

James A. Schafer Jr., *The Business of Private Medical Practice: Doctors, Specialization, and Urban Change in Philadelphia, 1900–1940*

David G. Schuster, *Neurasthenic Nation: America's Search for Health, Happiness, and Comfort, 1869–1920*

Karen Seccombe and Kim A. Hoffman, *Just Don't Get Sick: Access to Health Care in the Aftermath of Welfare Reform*

Leo B. Slater, *War and Disease: Biomedical Research on Malaria in the Twentieth Century*

Paige Hall Smith, Bernice L. Hausman, and Miriam Labbok, *Beyond Health, Beyond Choice: Breastfeeding Constraints and Realities*

Matthew Smith, *An Alternative History of Hyperactivity: Food Additives and the Feingold Diet*

Rosemary A. Stevens, Charles E. Rosenberg, and Lawton R. Burns, eds., *History and Health Policy in the United States: Putting the Past Back In*

Barbra Mann Wall, *American Catholic Hospitals: A Century of Changing Markets and Missions*

Frances Ward, *The Door of Last Resort: Memoirs of a Nurse Practitioner*

CPSIA information can be obtained
at www.ICGtesting.com
Printed in the USA
LVOW01*1711300117
522619LV00012B/168/P

9 780813 574394